Animals of Asia *the ecology of the Oriental Region*

Wild water buffalo can still be found in parts of northern India. Their longer horns distinguish them from working water buffalo and from feral animals that have escaped from domestic stock.

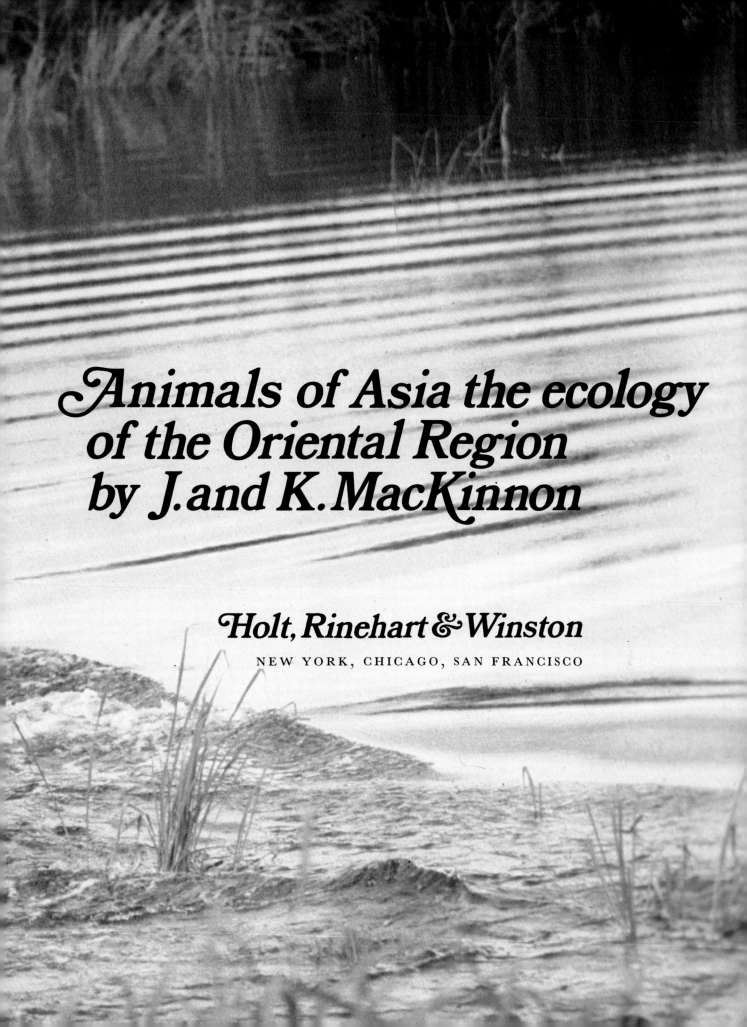

Animals of Asia the ecology of the Oriental Region
by J. and K. MacKinnon

Holt, Rinehart & Winston
NEW YORK, CHICAGO, SAN FRANCISCO

Introduction

Index compiled by M. O'Hanlon

Published simultaneously in Canada by
Holt, Rinehart and Winston of Canada, Limited.

Library of Congress Cataloging in Publication Data
MacKinnon, John Ramsay.
  Animals of Asia

  1. Zoology – Asia. 2. Zoology – Ecology.
I. MacKinnon, Kathleen, joint author. II. Title.
QL300.M3 1975   591.9'5   74-19372
ISBN 0-03-014116-8

Printed in Spain by Mateu-Cromo Artes Graficas S.A.
First American edition

Previous page:
Spinifex grass tolerates sea water, and is the first colonizer of
sand ridges on the sea shore. Later, the ridges will form new areas
of forested land.

Introduction

In 1271 young Marco Polo set out from Venice on an incredible overland journey to Cathay. He came back to Europe by way of Indo-China, Sumatra and India. The journey took sixteen years and on his return he wrote his 'Book of Marvels'. It was a strange tale of vast plains, high mountain ranges and great forested islands. He described the wondrous civilization and cities of ancient China; the power of the mighty Kublai Khan; and many of the curious animals of the Orient, the long-horned sheep of the steppes, enormous elephants and rare rhinoceros.

Later explorers probed the sea routes to the Spice Islands of the East. Vasco da Gama reached India in 1497 and Magellan's fleet visited Borneo and the Philippines on their round-the-world voyage a quarter of a century later. It was an amazing new world they saw. People of all shades and all religions, living and thinking in ways so foreign to Europe; ascetic yogas, snake charmers, rich Moslem rajahs and simple peasants. During the nineteenth century the great naturalist-explorers, Wallace, Beccari, Hornaday and many others, began a more scientific description of the wild plants and animals of tropical Asia but their discoveries added to rather than detracted from the wonders of the region.

Today a jet flight from Europe to Singapore takes only a few hours, but the Orient has lost none of its excitement or atmosphere with easier access and greater familiarity. The pace of life, the hot sultry climate, the brilliant colours and the tantalizing smells are all so new to the western visitor.

The animals of tropical Asia are no anticlimax. This region is truly rich in exotic life forms that come up to everyone's expectation of the mysterious East. This is the home of the strange, man-like ape, the orang-utan, the big-eyed tarsier, amazing gliding mammals and reptiles, fish that climb trees and others that shoot down insects from overhanging branches, naked bats, birds whose nests are edible, giant snakes, dazzling peacocks and colourful pheasants. In addition to these marvels there is always the exciting possibility that the great jungles and inaccessible mountain wastes may hide creatures as yet unknown to man. Ever-hopeful expeditions are still mounted to search for the Himalayan *yeti* and the *orang-pendek* or short folk of South-East Asia. Undoubtedly there are new species to be discovered in the remoter regions but whether these will prove as exciting as the mythical animals of native folk lore only time will reveal.

In this book we have attempted to present a picture of the ecology of probably the most diverse and fascinating of all the zoogeographic zones, the Oriental region. Only by considering the geological origins and climate of Asia can we understand the present-day distribution of habitats and their associated fauna. We have tried to show how each species is adapted for life within its environment and to explain its special relationships with its animal neighbours, whether as predator or prey or merely as competitors for food and space in the intricate web of life.

Human influence on the ecology of the Oriental region merits discussion in a chapter of its own. Man has had a long and active history in tropical Asia and the relationships between Man and wildlife run deep. Some animals are regarded as sacred, some have been domesticated, others are hunted for food or sport. Even more dramatic, however, are the changes wrought on the vegetation and climate as a result of agricultural practice and it is by such alteration of the environment that Man has had his most serious effect on the fauna.

The ecology of tropical environments is extremely complex. To understand fully the relationships between even a single species and its environment may take more than a lifetime of hard fieldwork. Our comprehension of the way in which whole forest or savanna ecosystems are controlled is still in a very elementary stage. A vast amount of research remains to be done before we can answer many of the problems we know exist; but our knowledge will never be complete for not only is the subject too vast, time is too short. Every year the lowland forests are shrinking before the saw and the axe and soon there will be no large tracts left unspoilt. And as their habitats vanish so, too, do the animals.

The chances of conserving the natural fauna of tropical Asia depend entirely on the amount of widespread interest that these animals can arouse. We hope that through the pages of this book the reader will come to share some of our own fascination and delight in the wildlife of the Oriental region.

John and Kathleen MacKinnon

Asia and the Oriental Region

sun be[ar]

elephant

sambar

hyena

rhinoceros

tapir

In its wealth of plant and animal life, in numbers as well as varieties, tropical Asia is the richest corner of our planet. The Oriental region, which stretches five thousand miles along the warm belly of Asia, is one of the six zoogeographic regions into which P. L. Sclater divided the world, according to the distribution of different bird species. It says much for his knowledge and insight, and for the reality of his concepts, that his regions are still accepted almost unchanged, each supporting its own characteristic fauna. They correspond approximately, but not exactly, to the five continental masses of North America, South America, Africa, Australia and Eurasia, with the last divided into two regions, the Palaearctic to the north and the Oriental to the south-east.

The Oriental region may be roughly defined as tropical Asia and its associated continental islands. It is an area of enormous complexity and habitat diversity. Rising from the sea to the highest peaks on Earth and from the region of heaviest rainfall to the arid heat of the Thar desert, it is also the home of nearly half of the world's human population. Man has been active here for over a million years. He has changed the face of Asia, changed its flora, altered its climate and greatly affected its wildlife.

Change, however, is not new to this region, for the world itself is in a constant state of flux. At present we are living in the middle of an unstable era, an Ice Age interglacial, the warmer period between two spells of severe glaciation. Several times during the last million years the polar ice sheets have enlarged and expanded. Sea levels have dropped and glaciers have scoured the temperate zones and tropical mountains. Life forms, too, are far from stable. As new varieties evolve, natural selection favours those which are better adapted to the changing environment and pre-existing types are superseded by new models.

Even the continents cannot be taken for granted as permanent. They lie on huge plates of the earth's crust, balanced on the viscous mantle. Turbulent convection currents from deep within our planet move these plates about the earth's surface, like floating leaves on a hot cup of tea. East Africa is tearing free along the Great Rift Valley. The Atlantic broadens five centimetres a year. Australia surges north and, if it continues at its present rate, within fifty million years it will have crashed through most of Indonesia. Where two plates move apart liquid rock wells up to seal the gap, adding new material to each, as is now happening along the Mid-Atlantic Ridge. Where two plates collide one subducts or dives beneath the other, melts and becomes incorporated into the fluid mantle. This is taking place along the line of the Andes, where the Pacific floor plunges beneath South America.

In the four and a half billion years since the earth formed, the continents have changed position many times. Sometimes they drifted slowly together, merging to form super-continents, at others they floated apart. Just 250 million years ago the continents were bunched together in one single great land mass now referred to as *Pangaea* (all land), surrounded by ocean, *Panthalassa* (all sea).

During the Triassic (about 200 million years ago), the great super-continent began to break up. The northern landmass, Laurasia, which was eventually to become North America, Europe and northern Asia, split off from the southern mass, Gondwanaland. Animal life was at an early stage of development. Birds had not yet evolved and the first primitive mammals were only just appearing – small, insignificant animals with sensitive facial vibrissae, woolly coats and sweat glands to regulate their body temperature. Specialized sweat glands produced sweet, nourishing secretions to feed the young, which probably still hatched from eggs like those of the platypus today. Early turtles, plesiosaurs and carnivorous ichthyosaurs flourished in the tropical seas but the dominant land animals were the mammal-like reptiles. They were a diverse group, including the rodent-like *Bienotherium*, carnivorous *Cynognathus* and the abundant *Lystrosaurus*, whose remains have been found in South America, Antarctica and India, adding biological confirmation to the geological theory of Pangaea. *Lystrosaurus* was 90 to 120 centimetres long and its elevated nostrils suggest that it was aquatic in habit, perhaps very similar to the hippopotamus today.

During the Jurassic (195 to 136 million years ago), Gondwanaland split into three large pieces: one which was to become South America, Africa

Previous page: Some of the animals inhabiting Java during the Pleistocene. Many of the species then living there now have a more limited distribution in the Oriental region.

It is now well established that the present continents were formerly a single landmass which started to break up about 200 million years ago. Subsequent continental drift took place and has produced the present-day distribution of the lands and oceans. The ancient landmass has been named Pangaea, divided into a northern area, Laurasia and a southern, Gondwanaland.

The table correlates the progress of drift with the geological periods and with the evolution of reptiles, birds and mammals. The four maps illustrate this progress, with special emphasis on the spectacular northward migration of the Indian sub-continent.

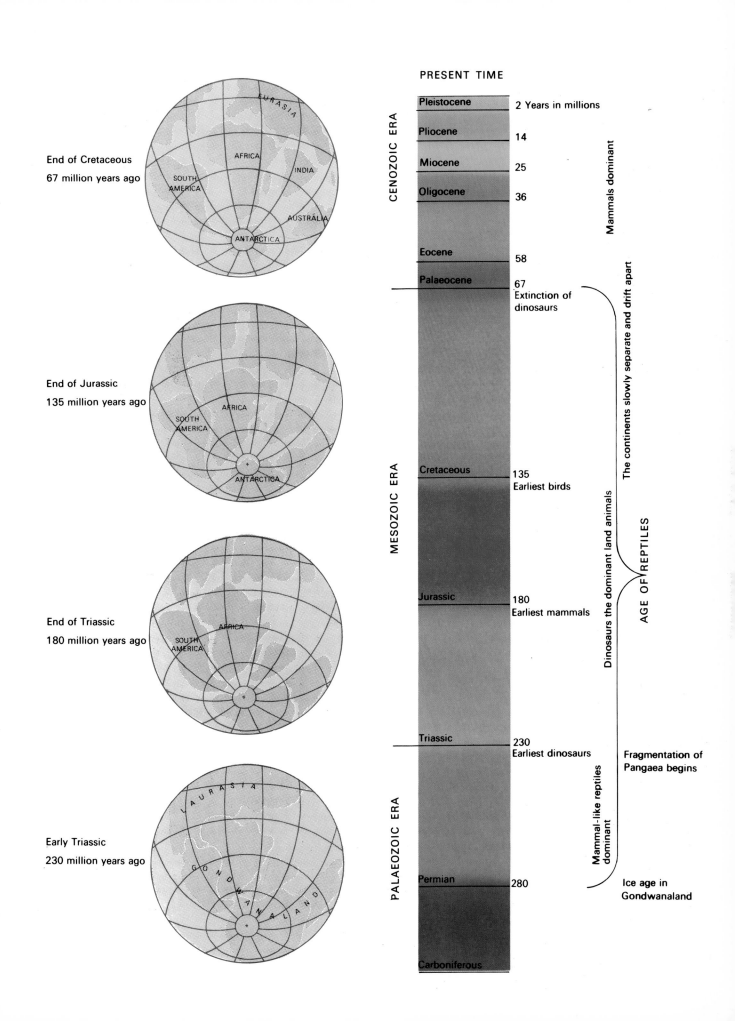

End of Cretaceous
67 million years ago

End of Jurassic
135 million years ago

End of Triassic
180 million years ago

Early Triassic
230 million years ago

PRESENT TIME

CENOZOIC ERA	Pleistocene	2 Years in millions
	Pliocene	14
	Miocene	25
	Oligocene	36
	Eocene	58
	Palaeocene	67

Mammals dominant

Extinction of dinosaurs

MESOZOIC ERA

Cretaceous — 135
Earliest birds

Jurassic — 180
Earliest mammals

Triassic — 230
Earliest dinosaurs

The continents slowly separate and drift apart

Dinosaurs the dominant land animals

AGE OF REPTILES

PALAEOZOIC ERA

Permian — 280

Mammal-like reptiles dominant

Fragmentation of Pangaea begins

Ice age in Gondwanaland

Carboniferous

and Madagascar, a second which was to produce Antarctica and Australia and a third, smaller fragment which drifted northwards, eventually to become the peninsula of India. Throughout this period the mammal-like reptiles were being replaced by a great radiation of different cold-blooded dinosaur types, not merely the familiar giant forms whose skeletons tower over the forecourts of the world's museums but an enormous variety, large and small, vegetarian and carnivore, quadrupeds, bipeds and gliders. By the late Jurassic the first birds had developed the pinnate feather, the secret of their later success, enabling them to fly as well as regulate body temperature. Several lines of small mammals continued to evolve.

Throughout the Cretaceous (136 to 65 million years ago) the land masses continued to re-align. South America split from Africa, which swung round to join Eurasia. India was still drifting north and Antarctica south. By the closing stages of the Mesozoic era the mammals had established four main lines: the egg-laying monotremes and three groups producing live young, the Multituberculata (an aberrant group which later became extinct), the pouched marsupials and, most successful of all, the placental mammals or Eutheria. Birds, too, were well established but the reptiles, especially the dinosaurs, still dominated the land and sea.

Suddenly, with the end of the Cretaceous, some 70 million years ago, everything changed. Three-quarters of all the cold-blooded reptile families became extinct. All the dinosaurs, pterosaurs, ichthyosaurs and plesiosaurs died out. A radiation of warm-blooded mammals and bird-types filled most of the faunal vacancies and the mammals have been the dominant land animals ever since. Many theories have been advanced to explain the end of the Age of Reptiles – cataclysmic events not recorded in our rock history, unusual solar radiation, widespread epidemics, the evolution of egg-eating mammals. Though these remain possibilities, it is far more likely that the extinctions occurred gradually over hundreds of thousands of years. This was a period of climatic and geographical changes. The world temperature began to drop and more temperate climates prevailed. As the continents realigned, great mountain ranges upfolded and, as Africa and Laurasia drew together, species interchange became possible. In any faunal reshuffle one would expect the more advanced and warm-blooded animals to expand at the expense of the dinosaurs, which were ill adapted to change.

During the last sixty-five million years North

MACKINNON

America has separated from Europe, and Australia from Antarctica; India has crashed into southern Asia, causing the Himalayan range to rise up like a rumpled rug. Until this happened India may well have lacked any mammals and may have been reached by only a few birds but it now became linked by land with the rich fauna of North Africa. One of the first mammals to move into this newly available environment was of enormous size: fossil remains of *Indricotherium*, a huge, hornless rhinoceros which stood fully 5 metres high have been found in Oligocene deposits in Russia, Mongolia, Pakistan and East Africa.

By late Caenozoic times the Indian fauna was well established. Pliocene deposits in the Siwalik Hills of north Punjab have yielded an accurate picture of the animal life of twelve million years ago. Among the giant creatures existing then were *Sivatherium*, a branch-horned giraffe, and *Deinotherium*, a curious elephant, bearing downward-curving tusks on the lower jaw. It is difficult to imagine what function these tusks could have had. The presence of *Hipparion*, a primitive, three-toed horse, together with aardvarks, hyaenas and antelopes, indicate that the area was covered by savanna and woodland.

A long belt of volcanic activity runs through Java and Sumatra, producing very fertile soil, but also the hazard of earthquakes and eruptions. This desolate scene near Sibajak in Sumatra shows a semi-active crater in the foreground, with jets of steam and deposits of sulphur. Further away, dense rain-forest stretches to a volcanic cone.

The primate fauna at this time is especially interesting. It included macaque monkeys and three species of early ape: *Dryopithecus indicus*, a large, gorilla-like creature which gave rise to the extinct *Gigantopithecus* of Pleistocene China, *D. sivalensis*, the probable ancestor of the orang-utan, and a pygmy form, *D. chenjiensis*, which seems to have left no descendants. Most exciting of all the Siwalik discoveries are a few jaw fragments of *Ramapithecus*, a very primitive hominid ape belonging to Man's own family. For some years archaeologists have thought that *Ramapithecus* may have been directly ancestral to Man, but more complete remains of the animal from East Africa suggest that it may be only another curious sideline of ape evolution.

The Sunda Shelf, stretching from Indo-China, through Borneo and Sumatra to Java, is a comparatively recent addition to the Oriental region. It was not until the end of the Miocene (15 million years ago) that the first small islands emerged from the sea, where the western Javan Plateau is situated today. During the Pliocene (15 to 1 million years ago) widespread tectonic movements caused more land to rise up on Java and Borneo and an attendant period of volcanic activity resulted in the eruption of two main belts of volcanoes, one run-

ning from north Celebes through the Philippines and the other, longer chain stretching from the Lesser Sunda Islands, through Java and along the length of Sumatra. Volcanoes dominate the Javanese and Sumatran landscapes and the area is still violently unstable. In 1883, the region was rocked by the loudest explosion recorded in historical times when the tiny island of Krakatoa, midway between Java and Sumatra, disintegrated. The sound of the blast was recorded as far away as Ceylon, the Philippines and Western Australia. Huge tidal waves destroyed hundreds of coastal settlements, killing more than 40,000 people. Air waves, caused by the explosion, passed three times round the world and fine particles blasted into the upper atmosphere produced remarkable coloured sunsets over the whole globe for more than two years. A new volcanic island, Anak Krakatoa (son of Krakatoa), has now emerged in its place. This is still active, rumbling and smoking and tossing out great boulders, some as large as a house, high above its rim.

The Pleistocene (the last million years) was a period of world-wide instability. Dramatic fluctuations in climate produced a succession of glaciations and interglacials in temperate regions and pluvials and interpluvials in the tropics. Although climatic changes in South-East Asia were less marked than elsewhere during this period, they were still of paramount importance in determining the fauna of the Sunda region. With so much water frozen in the polar ice caps during the glaciations, sea levels throughout the world dropped and the islands of Borneo, Java and Sumatra were connected by land bridges to the mainland of Asia. Along these bridges successive waves of animal migrants were able, over a period of half a million years, to reach the islands

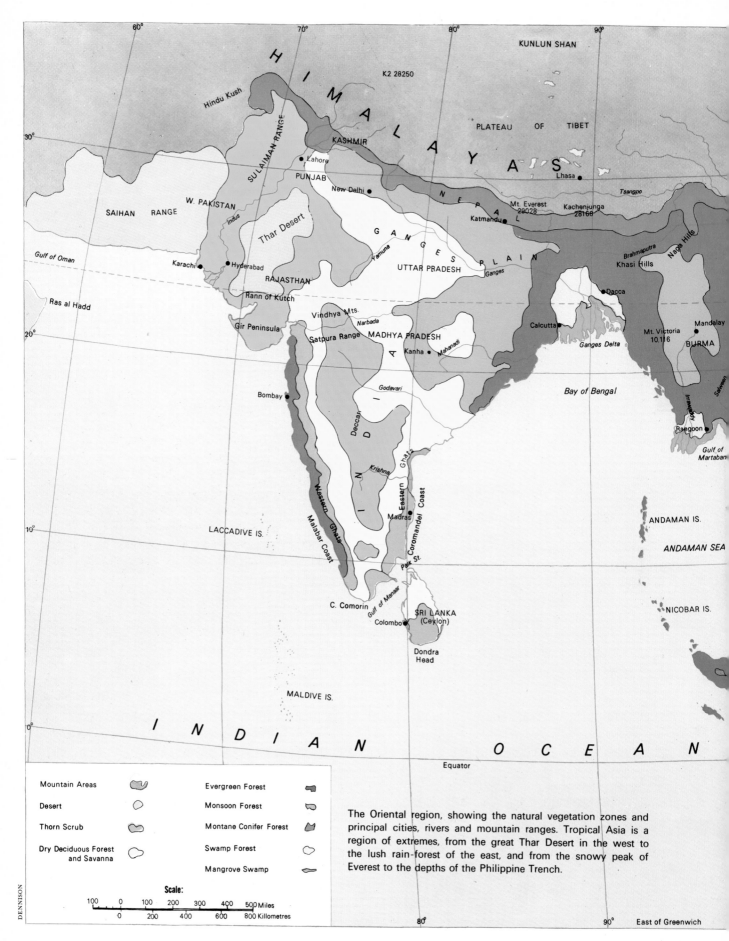

The Oriental region, showing the natural vegetation zones and principal cities, rivers and mountain ranges. Tropical Asia is a region of extremes, from the great Thar Desert in the west to the lush rain-forest of the east, and from the snowy peak of Everest to the depths of the Philippine Trench.

Legend:

- Mountain Areas
- Desert
- Thorn Scrub
- Dry Deciduous Forest and Savanna
- Evergreen Forest
- Monsoon Forest
- Montane Conifer Forest
- Swamp Forest
- Mangrove Swamp

Scale:

| 100 | 0 | 100 | 200 | 300 | 400 | 500 Miles |
| 0 | 200 | 400 | 600 | 800 Killometres |

DENNISON

18

East of Greenwich

of the Sunda Shelf. In the warmer interglacial periods, the melting ice caps receded and sea levels rose again, submerging the land connections so that the islands were isolated once more.

Fossil evidence shows that Java supported a rich Pleistocene fauna. As well as species still found there today, and their ancestral forms, it included bears, hyenas, elephants, tapirs, antelopes, scaly pangolins, siamangs and orang-utans – animals whose distribution in the Oriental region is now much more limited. Two species of Early Man also occurred in Java: *Meganthropus* and Dubois' famous discovery of *Homo (Pithecanthropus) erectus* are the oldest known remains of Man in Asia.

Wallacea and the boundaries of the Oriental region

Geologically Java and the adjoining island of Bali lie at the eastern limit of the Oriental region. If a submarine contour is drawn at a hundred fathoms, following the coast west of Burma round to east of Vietnam, it will enclose an area including the Malay Peninsula (henceforth referred to as Malaya), Sumatra, Borneo, Java, Bali and Palawan. This contour marks the edge of the continental or Sunda Shelf and includes the limits of all land that was connected to the mainland during the Ice Ages when the sea level fell by over 100 metres. A contour map of the Sunda Shelf shows quite clearly the great submerged river system of the ancient Sundaland, draining a deep trench between Sumatra and Malaya to the west and Borneo to the east. The formerly connected tributaries can still be recognized by the species they share. The Kapuas River in west Borneo has fish almost identical to those found in the rivers of eastern Sumatra but quite different from those of the Mahakam river in east Borneo. Similarly the freshwater and estuarine crustaceans of eastern Johore in Malaya are quite different from those on the Malayan west coast but include a number of species which are also found on the coast of Sarawak.

East of the continental shelf lie Timor and the Lesser Sunda Islands, Celebes, the Moluccas and the Philippines. These have arisen out of the sea, largely as a result of volcanic activity, and have never been connected to the Asian mainland, even at the times of lowest sea level. Nor do they lie on the Australasian continental shelf as does New Guinea. This archipelago between the two continental shelves is sometimes referred to as Wallacea and zoogeographically is perhaps the most interest-

ing region in the world. The fauna, though impoverished in species, is a mixture of both Oriental and Australasian forms; the relationships of the different islands' wildlife with each of these major regions has been the subject of much debate.

The famous nineteenth-century explorer and naturalist Alfred Russel Wallace, after whom the area is named, was the first person to recognize the sharp faunal break that occurs to the east of the islands of Bali and Borneo. He proposed a faunal limit running from between Bali and Lombok in the south, northwards through the Macassar Strait and including the Philippines on the Asian side of the boundary. T. H. Huxley later modified this to exclude all the Philippines except the island of Palawan and he proposed the name 'Wallace's Line' for this faunal boundary which now precisely followed the limits of the Sunda Shelf.

Across the rest of the archipelago there is a gradual change in fauna. Early this century the Dutch zoologist Max Weber suggested another line of faunal balance to distinguish between those islands whose animals showed affinities with the Oriental region and those which had fauna of Australasian origins. Weber's Line, based mainly on the distribution of mammal and mollusc species, lies further east than Wallace's Line. It places the Philippines, Celebes and Timor within the Oriental region, but the Moluccas, Buru and Tanimbar, islands lying off the western margin of the Sahul Shelf, on the Australasian side.

Of the debatable land included by Weber within the Oriental region, Celebes has the most interesting fauna. Most of its mammals, including pigs, deer, civets, squirrels and a tarsier are of obvious Asian origin. Rats and mice occur in both regions and could have arrived from either. There are several species of mammals endemic to Celebes. Two of these, different types of cuscus, are marsupial phalangers, typically Australasian. They are strange, arboreal animals with strong prehensile tails. Other endemic species show Oriental affinities. They include the anoa, smallest of all known oxen; the babirusa, an aberrant pig with long, curved tusks of unknown function; and two species of monkey, the moorish macaque and the so-called black ape. The latter is a fierce-looking, long-faced monkey, which shows many similarities to the African baboons. Black apes are quite tail-less and march about the forest in large troops or jump skilfully among the treetops, where they feed on fruits and young shoots.

Celebes is richer in bird than mammal species,

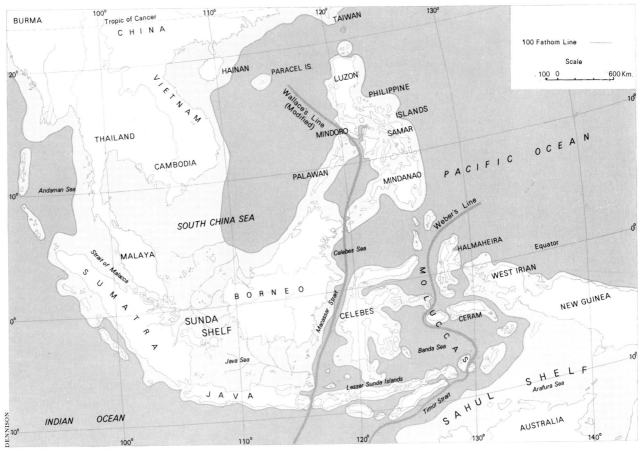

Map of Wallacea showing the islands of the Sunda and Sahul continental shelves and the 100 fathom contour. This contour corresponds with the extent of land exposed at the time of lowest sea level during the Pleistocene. At that time animals could travel overland from Asia to all parts of the Sunda Shelf, the boundaries of which are delineated by Wallace's line. Islands between the two continental shelves, Celebes and the Lesser Sunda Islands, have derived their fauna from both the Oriental and Australasian regions. Weber's line marks the position of faunal balance between the two regions.

The island of Celebes is part of the zoologically debatable land between the Oriental and Australasian regions. Most of its mammals are clearly of Asian origin, but the Celebes cuscus is a marsupial with Australian affinities which has evolved into a distinct species on the island. The cuscus is arboreal, climbing with all four limbs and a grasping, prehensile tail.

which is to be expected since it is easier for flying birds to colonize an oceanic island than it is for animal migrants dependent on floating vegetation for transport. The avifauna is made up mostly of Oriental families – hornbills, doves, drongos, king-fishers, sunbirds, crows and herons. The fascinating maleo fowl, however, is a member of the Austral-asian brush turkey family (Megapoda). These birds feed on fruits throughout the forest but congregate at breeding times on beaches of volcanic ash. Here the female maleo fowl dig tunnels more than a metre deep where they lay their large, red eggs. To facilitate this tunnelling the female bears a special knob on her head. The eggs are incubated by the natural heat of the ground and, when they hatch, the young emerge fully-fledged and able to live independently.

The Philippine Islands, too, show strong Oriental affinities. Mammals include the tamarou, which is a dwarf buffalo like the anoa, a flying lemur, a tar-sier, monkeys and rats. Birds are similar to those of Celebes. In addition, this is the home of the spectacular monkey-eating eagle which, sadly, has been hunted to the verge of extinction by over-enthusiastic taxidermists.

Island-hopping mammals have been able to

The so-called 'Komodo dragon' is the largest of all the monitor lizards, being longer and bulkier than any other species. It grows to over 4 metres in length and about 140 kilograms in weight. It is found only on a few small islands of Indonesia, with the largest population on Komodo.

The magnificent monkey-eating eagle is one of the largest of all birds of prey. It hunts monkeys in the treetops of rain-forest in the Philippine Islands, but, sadly, it has now been almost exterminated by trophy hunters.

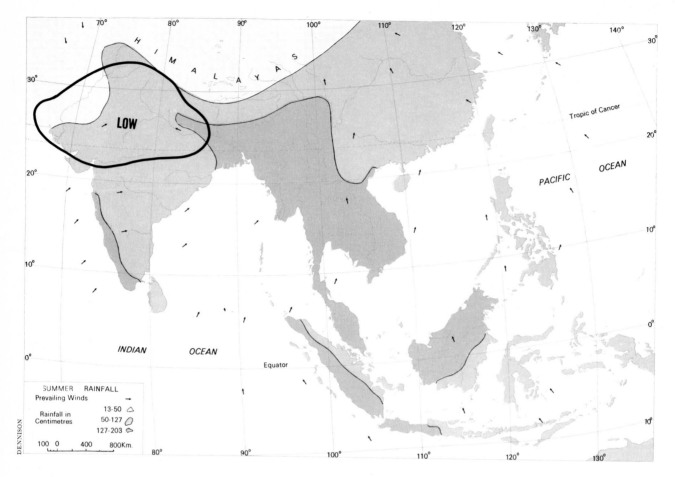

spread down the island chain to reach the Philippines and Celebes from southern China via Formosa. Celebes also lies very close to Borneo and during the Pleistocene the straits between the two islands were probably no more than 50 kilometres wide. Even so, of the eighty species of terrestrial mammals and over sixty species of bat which occur on Borneo, less than a tenth have reached Celebes. Even fewer land mammals have been able to reach the Timor group. Five of these, the long-tailed macaque monkey, a pig, a deer, a civet and a shrew, are of obviously Oriental origin; only the cuscus is Australasian. Birds and insects of the group are predominantly Oriental, although Timor itself boasts many Australasian species. The great Komodo monitor lizards, over 4 metres in length, have put the small islands of Komodo and Flores on the zoological map. But since monitors are a group of reptiles which occur in both the Australasian and Oriental regions, their presence here tells us little about the area's true faunal relationships.

The other boundaries of the Oriental region are less difficult to define, although, with the exception of the southern shoreline along the Indian Ocean, they are by no means precise. The region has been defined as essentially tropical Asia and its closely

associated continental islands, but it is more satisfactory to relate the boundaries to geographical features. We will, therefore, take the northern boundary, with the Palaearctic, from the Hindu Kush mountains in the west, eastward along the Himalayan massif to include Yunan, Szechuan and the island of Taiwan (Formosa). The western boundary, again with the Palaearctic, lies on the western side of the Indus valley, excluding Baluchistan and Afghanistan. This vast area includes a great variety of habitats, each supporting its own distinctive fauna and it is convenient when looking at the relationships between the animals to recognize three subregions: the Indo-Chinese, the Indian and the Indo-Malayan.

The Indo-Chinese subregion, including the Himalayan mountain system, has a fauna showing strong affinities with the Palaearctic. The Himalayan range is a comparatively recent geological phenomenon, not always the barrier to animal migrations that it is today. Bears and wild pigs, tigers and snow leopards are all found on both sides of the Himalayas, in the Palaearctic as well as the Oriental region.

The Indian subregion was once linked by continuous woodland savanna with the Ethiopian

24

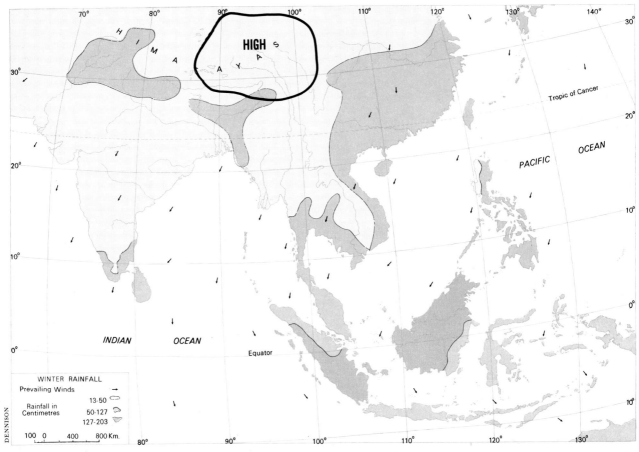

Two monsoons bring rain to tropical Asia. In summer (left) a belt of low pressure builds up over the Asian mainland and rain-bearing winds blow north from Australia and the Indian Ocean. In winter (above) there is an area of high pressure over central Asia but an area of low pressure in Australia. Winds now blow south and east with little rain falling on the mainland.

region of North Africa. Desert barriers now separate the two areas but the faunas remain very similar, sharing such animals as lion, cheetah, leopard, hyaena and antelopes.

The Indo-Malayan subregion has retained its natural rain-forest vegetation since the Pliocene. While it has received its animals from the two subregions mentioned above, they have evolved within this stable habitat to give the fauna of the area a distinctive character. Within this rich and complex environment there exists a wealth of animal types. Species such as the gibbons, orang-utan, tarsiers, tree shrews, tapirs and flying lemurs are confined almost exclusively to this region.

Climate and vegetation

Three major physical factors determine the distribution of plants and animals throughout the Oriental region. These are altitude, latitude and rainfall. Increases in altitude and latitude result in decreases in temperature. For every 1,000 metres height, or ten degree shift in latitude, the temperature drops about 5°C. Thus, as one moves north, tropical forms give way to more temperate species. In the case of the Oriental region there is a broad transition zone in southern China, drawn not far above the Tropic of Cancer, where many tropical groups reach their northern limit. At increasing altitudes the number of plant and animal species declines. To cope with the extreme environment they must resort to physical and physiological adaptations. Plants tend to decrease in size and produce stunted alpine forms. Mammals overcome the heat problem by growing thick, shaggy coats but their distribution is still limited by food availability.

Rainfall must be considered not only in terms of annual total but also in its distribution throughout the year. Marked seasonality of rainfall determines the type of plants able to live in any region. Equatorial lowland areas which receive at least 6 centimetres of rain each month can support evergreen rain-forest. Areas with marked dry seasons can support deciduous or monsoon forest only if the total annual rainfall is very high. When rain falls more evenly throughout the year, the total amount needed to support forest vegetation is less. Where rainfall is too low, or too seasonal, savanna, scrub or even desert are found.

To understand the distribution of vegetation types throughout the Oriental region the prevailing weather must be considered. Rain in tropical Asia arrives during a summer monsoon and a winter monsoon. In summer the path of the sun is directly over northern India and southern China. As the air warms up, its molecules become more dispersed, causing an area of low pressure over central Asia. Monsoon winds blow up from the high pressure zones now existing over Australia and the Indian Ocean. As they pass over wide expanses of sea they pick up moisture. When these winds reach the Asian land mass they are forced to climb to cross hills and mountains and, as they climb, they cool and drop their moisture. Thus heavy monsoon rains fall on India and south China and more moderate rains fall on the Sunda Shelf, the Philippines and New Guinea. Australia's great bulk, however, causes a rainshadow which results in very little rain falling on the Lesser Sunda Islands, south Celebes and eastern Java. This is reflected in the seasonal vegetation of these areas: in the western end of Java, which is not affected by the Australian rainshadow, we find luxuriant rain-forest while 700 kilometres away, on the east coast, there is savanna, scrub and cactus.

In winter central Asia is very cold, producing a high pressure zone. The monsoon winds now blow southwards towards the equator and the hot, low pressure zones now found over Australia. They are met by other winds blowing up towards the equator from the high pressure zone over the southern Indian Ocean. Where the hot and cold air meet heavy rain falls over all of the Sunda Shelf, Celebes, the Lesser Sunda Islands, New Guinea and north-eastern Australia.

Those regions receiving regular heavy rainfall, and thus able to support evergreen rain-forest, include the Western Ghats of India, parts of Ceylon and most of Burma, Thailand, south China, Malaya and the Greater Sunda Islands. The natural vegetation of the Indian peninsula is mainly monsoon and other types of deciduous forest and thorn scrub. In the drier north-east corner lies the Thar desert, on the Indo-Pakistan border.

While maps of natural vegetation show a close correlation with rainfall, they fail to give an accurate picture of the existing vegetation today. This is especially true for peninsular India and the more populous part of South-East Asia where forest clearing for timber, animal pasture and agriculture has been extensive and is continuing at a frightening rate. As can be seen from a map of modern land usage only 20 per cent of India still supports natural forest and the consequent decline of many Indian wild game species is inevitable.

Right: The type of vegetation that a given area can support is determined not only by the total rainfall, but by the way that rainfall is distributed throughout the year. This diagram shows the approximate relationship between vegetation, total annual rainfall and the rainfall during the driest month of the year for equatorial lowland regions.

Example:
Total annual rainfall 200 cms.
Rainfall in driest month 8 cms.
Vegetation: evergreen rain-forest

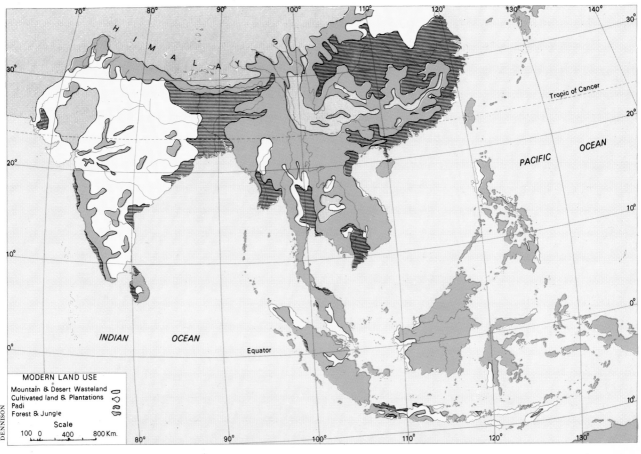

Modern land use in tropical Asia. Almost half the world population lives in this area. In the most densely peopled parts, India, China and Java, much of the natural vegetation has been cleared for agriculture. Less than 20 per cent of India and 10 per cent of China still supports natural forest.

MODERN LAND USE
Mountain & Desert Wasteland
Cultivated land & Plantations
Padi
Forest & Jungle

Scale
100 0   400   800 Km.

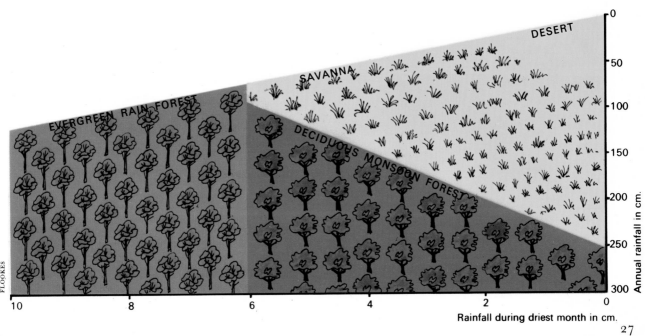

The Tropical Rain-Forest

Asian elephants are found in many parts of Asia where forest survives. They frequent clearings and river courses where there is plenty of ground level vegetation, and often invade plantations and village gardens. They are very destructive feeders and where they occur at high densities they seriously affect the growth pattern of the forest.

Tropical rain-forest contains more different species of tree than any other type of forest. In Bornean rain-forest 3,000 species have been recognized, and well over 100 of these can be found in any selected hectare.

The rain-forests of the Oriental region offer the most favourable environment for life on Earth. The equatorial climate, with constant high humidity and high temperatures, combined with heavy year-round rainfall and plentiful strong sunlight provides perfect conditions for maximum plant growth. Since the climate is stable throughout the year the trees show no seasonality in their fruiting or flowering. Different tree species, or even the same species in different valleys, may produce new leaf growth or crop at different times of the year so that food specialists among the animals can always find their preferred foods, even if they have to range widely to do so. Moreover, not only does this region support some of the tallest, densest, most luxuriant forest in the world, it can boast a remarkable number of plant types. In Borneo alone there are some three thousand species of trees, not to mention the mosses, ferns, epiphytes, orchids, lianas and vines, herbs, shrubs and fungi that make up this rich and diverse habitat. A single hectare of jungle may include a hundred or more different tree types and the neighbouring hectare will add almost as many new species to the list.

Within this rich environment live a host of animals of all kinds: herbivores such as elephants, muntjacs and caterpillars; fruit-eaters such as squirrels, hornbills and fig wasps; and predators such as leopards, linsangs and spiders. To understand how these creatures fit into the ecology of the rain-forest, it is necessary to consider first the nature of the forest itself.

Giant trees stretch up towards the sun. Buttress roots anchored in the thin soil support the smooth, straight trunks, which rise 30 metres or more before branching out. The spreading crowns effectively block out most of the light from the shady layers beneath. The predominance of the many types of dipterocarp trees (named after their two-winged fruits) among the continuous leafy canopy has led botanists to describe this type of vegetation as dipterocarp forest.

Even taller trees tower above the canopy. These emergents are mainly leguminous trees, members of the pea family, which are at an advantage on the rain-leached soils because of their ability to fix nitrogen in their roots. The tallest of them all is the *tualang, Koompassia excelsa,* which may exceed 80 metres in height. It is on the branches of the *tualang* that the large wild bees of South-East Asia prefer to build their pendulous combs, where they are safe from the ravages of the small, black sun bears and sweet-toothed monkeys. They are not secure from Man, however, for the nerveless Dyaks construct flimsy sapling ladders, peg them to the giant trunks and, at considerable risk, collect the precious honeycombs.

A multitude of creepers, climbers, vines and lianas take advantage of the tree supports to enable them to expose a few leaves to the sunlight. One of the most interesting groups, which, because of its rich sweet fruit, is of particular importance to the local animal life, is the strangling figs. Taking root on a lofty branch the fig seedling drops down a slim, dangling root, which anchors on the forest floor to obtain the water and minerals that are vital for the plant's growth. Gradually the fig grows larger and, as it grows, it drops more twisting aerial roots, until these eventually strangle the original tree and the fig takes over the crown

The strangling fig grows first as an epiphyte, high in the branches of a tall tree. Later a hanging root reaches the ground, roots itself and the plant grows round its host squeezing and crushing it to death. Finally the host tree decays, leaving the fig tree standing.

Left: The tall trees of the rain forest may reach or exceed a height of 70 metres and must be very firmly rooted to withstand the force of the wind. Large, plate-like buttress roots increase the area of the tree's base and so improve its stability.

The stag's horn fern *Platycerium coronarium* is an epiphyte, growing on the branches of forest trees. It extracts no nourishment from the tree's living tissue but has roots that are adapted for collecting and storing rainwater from its host. It also benefits from being nearer to the sunlight. The stag's horn fern has leaves of two distinct kinds: 'nest leaves' which grow upwards and branched fertile leaves which bear the reproductive organs.

with its own productive leaves and shoots. The host tree gradually rots away.

Other smaller plants gain access to the life-giving daylight by clinging to the branches of the great trees. Unlike the strangling figs, these epiphytes have no contact with the ground. Yet they, too, must have water and minerals if they are to exploit the sunlight. To overcome these problems many have roots and tubers adapted for collecting and storing moisture from their hosts when there is rain. The stored water can then be used by the epiphyte during the dry midday heat. In this way the army of tropical orchids and ferns that festoon the lofty boughs are able to survive.

Minerals are obtained by even more curious methods. Pitcher plants have leaf traps which attract and catch insects. The victims are digested by an acidic enzyme in the water of the pitcher and provide the plant with essential nitrogen and salts. Other plants exploit ants. A plant inhabited by ants gains two major benefits. The presence of the ants deters potential leaf-eating animals and protects their host. Secondly, since the ants deposit their faeces in the hollows they occupy, the plant has a ready source of minerals. In return the ants are provided with a secure home, within leaf joints or tubers, where they can lay eggs and where their larvae can develop. Many types of plant-ant associations are found in the forest and some plants positively encourage ant 'lodgers' by secreting sweet liquids for the insects to feed on.

Working vertically down from the canopy, the rain-forest can be divided into three horizontal layers. The canopy and the emergents together make up the upper storey, which receives most of the available sunlight. Below this is the middle storey, composed partly of specialist trees, able to survive in the comparative shade, and partly of saplings of the taller dominant species. When one of the giant trees falls, leaving a gap in the canopy, the sunlight bursting through stimulates the saplings to surge up and take its place. The lower storey of the forest is almost bare. Very little direct sunlight ever reaches the forest floor and vegetation is sparse. Only a few seedlings, crouch palms, creeping rotan palms, liana roots and ginger plants bearing strange blossoms break the deep carpet of fallen leaves and decaying wood. It is only where the upper storeys are incomplete, along river-banks, where trees have fallen, or on hillsides too steep for large trees to gain a footing, that we find the impenetrable tangle of spiny vegetation that is the popular idea of a tropical jungle.

Amorphophallus prainii, an aroid plant common in lowland forests of Malaysia. Like its rare giant relative, the 2 metre high *Amorphophallus titanum*, it has both male and female flowers on the same stem but relies for cross-pollination on pollen-carrying beetles.

Tree ferns are among the most typical and beautiful of the small trees of the rain-forest. They look rather like palms but are true ferns, elevated on a woody stem. They are especially characteristic of wet, cloudy mountain forest around 1,000 to 1,500 metres.

The diversity of forest species

Why are there such a large number of different plant species within the rain-forest? One reason is, quite simply, that conditions are so favourable that almost any plant can grow there. The greatest tree diversity is found in the moist, flat valleys but going up into the hilly ridges the number of different species is sharply reduced. Few of the valley species can cope with the more demanding environment of drier, fast-draining soil, greater wind exposure, lower midday humidity and higher temperatures in the canopy. Of those species which do occur on the ridges most are also to be found on the valley floors but at a lower frequency.

Possibly the question should be reversed. Why is there no single species dominating the forest, as do oak and beech in more temperate regions? The answer probably lies in the speed with which plant-eating insects can evolve in the humid tropical forest. Their breeding continues uninterrupted so that several generations are born within the space of a year. Consequently if a characteristic evolves which benefits these insects it can spread through the whole population in a very short time.

To protect their leaves from being eaten by the multitude of forest insects, the trees have developed many types of poisonous and distasteful substances. At the same time the insects have been evolving a capacity either to tolerate or to break down these chemicals. If all plants used the same defence they would all be immediately vulnerable to any insect species that made a breakthrough. Hence the plants' best defence has been in diversity, each species developing new and unique chemical deterrents to limit the number of its potential predators. This has in turn led to diversification of the insects; different types have specialized in the breakdown of different chemical defences and have thus become associated with particular plant species. Not only has it been of advantage for a plant to be unique, but there has also been a premium on each species being rare. A species-specific insect pest is more likely to miss a single seedling than one in a clump of the same species.

The Oriental region has remained climatically stable for a much longer period than any temperate habitat. During the turbulent Pleistocene the rain-forest expanded and contracted several times but never vanished and the forest of today is very old indeed. The result of such a long selection for individuality and rareness among plants, and consequently their insect pests, is the amazing

Ginger is familiar to most people only as a flavour. It is in fact prepared from the roots of plants of the genus *Zinziber*, one of the few plants that bloom on the forest floor.

The feeding relationships in tropical rain-forest ecosystems are more complicated than those in any other habitat. This simplified diagram represents the way in which energy from the sun is accepted in the different forest strata and how the resulting plant energy is exploited directly or indirectly by the different groups of animals. All dead or discarded material falls to the forest floor where organic matter among the leaf litter is broken down by soil organisms. Vital nutrients are thus released to be absorbed through plant roots and recycled into the system.

diversity that survives in South-East Asia, where the lives of 15,000 species of plant and 150,000 species of animal are intricately interwoven.

Stratification and niche separation

The forest forms a complex three-dimensional matrix, whose structure must be taken into account when considering its animal inhabitants. Different parts of the forest offer quite different habitats to animals: different locomotor possibilities and different distributions of food and water, sleeping and breeding sites. Such physical factors as temperature, humidity and light incidence also determine where animals can survive, and these factors vary enormously between the forest layers.

As the forest plants can be described in horizontal

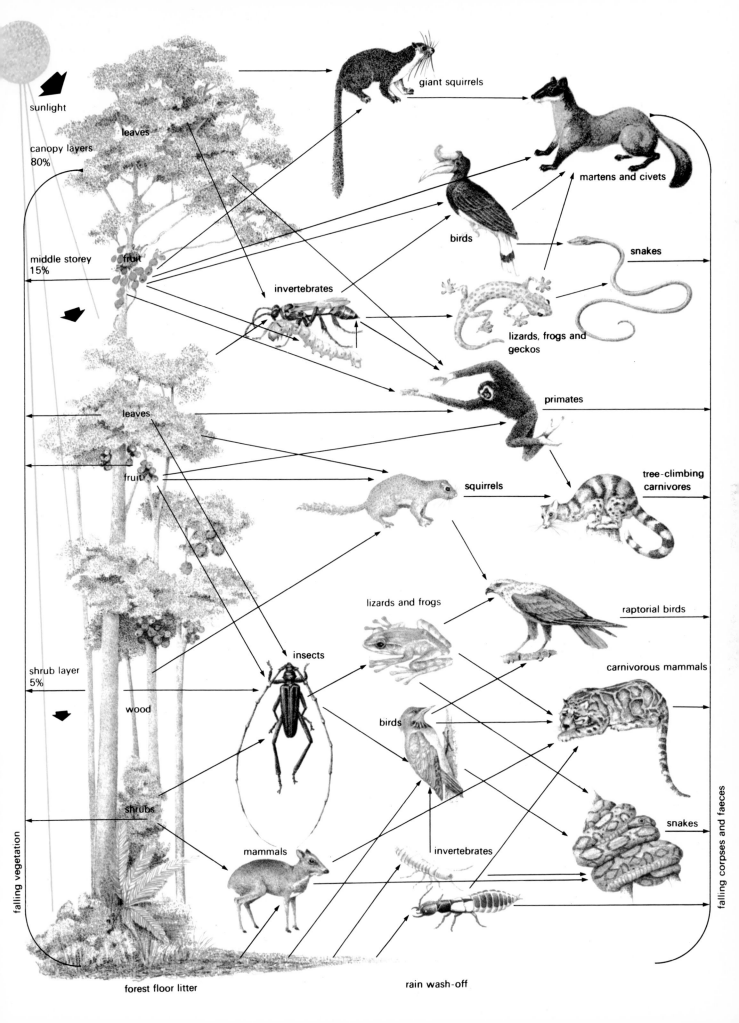

sunlight

leaves

canopy layers
80%

middle storey
15%

fruit

invertebrates

giant squirrels

martens and civets

birds

snakes

lizards, frogs and
geckos

primates

leaves

fruit

squirrels

tree-climbing
carnivores

lizards and frogs

raptorial birds

insects

carnivorous mammals

birds

shrub layer
5%

wood

shrubs

mammals

invertebrates

snakes

falling vegetation

forest floor litter

rain wash-off

falling corpses and faeces

Prevost's squirrel is the most striking of the many rain-forest squirrels. Its basic pattern is black above, chestnut red below, with a white stripe between the two on each flank, but it varies in colour regionally from mainly white to almost black. Though its home is in the forest, it adapts well to life in coconut and oil palm plantations.

Tree shrews are less arboreal than their name suggests. They spend most of their time probing among dead logs and under leaf litter searching for succulent insects. They are territorial and have specialized scent glands under the throat and in front of the groin with which they mark out their ranges. Borneo is the centre of distribution of all tree shrews: eleven out of seventeen known species are found there and eight are confined to the island.

strata, so the animals can be placed in corresponding categories: terrestrial animals ranging over the forest floor, arboreal creatures living high in the canopy and a wide range of intermediate types that seem equally at home on the ground or in the trees. Tropical rain-forests are generally characterized by the large number of arboreal animals they support.

One group which illustrates these principles very well is the squirrels. South-East Asia boasts dozens of species of squirrels of which some twenty-six occur in Malaya. At Bukit Lanjang, a fascinating long-term study has revealed their differential use of the forest. A horizontal catwalk has been erected from a steep hillside and suspended through the canopy above the valley floor. Its height ranges from 0 to 60 metres above the ground. Cage traps are set among the branches and lowered in baskets into the crowns of the middle storey trees below. Trapped animals are marked, examined and released and over the years a very good picture has emerged of how individual squirrel species exploit the various forest strata.

In the canopy there are two species of giant squirrel, *Ratufa*, both over 70 centimetres long, one with a striking black and white coat and the other

a rich creamy colour. Usually solitary by nature, these large rodents leave their leafy dreys in the early morning and range widely throughout the treetops, feeding on fruits and leaves. The small Prevost's squirrel is equally at home in the upper storey of the forest. This is one of the Callosciuridae or 'beautiful' squirrels and anyone who has seen it will appreciate that the group is aptly named. Prevost's squirrel has a glossy black back and tail, with a thick white stripe running along its sides, and legs and belly of chestnut red. Several are often to be seen feeding in a fig tree at the same time as a group of monkeys, apparently unperturbed by the bigger primates. At night the canopy is the domain of the flying squirrels, of which Malaya alone boasts eleven species. Like the 'non-flying' species, they range in size from giant forms, up to 80 centimetres long, to the pygmy flying squirrel, less than 20 centimetres long and so far found only in Selangor.

The middle layer of the forest provides a habitat for the rest of the 'beautiful' squirrels, including the common plantain and slender squirrels, which are equally at home outside the rain-forest in the artificial environment of rubber plantations or gardens, and the horse-tailed squirrel and its close cousins. All of these squirrels are scansorial, moving up and down the tree trunks in search of food. Since they supplement their vegetarian diet with small arthropods it is often worth their while to forage close to a monkey troop so that they can catch the insects disturbed by the monkeys.

Ground squirrels are by no means averse to climbing trees but they are usually seen scampering across the forest floor. The most common is the three-striped ground squirrel which lives, not in nests or holes as expected of squirrels, but in burrows in the forest floor. With three longitudinal black stripes on its back it looks not unlike a small American chipmunk. Ground squirrels take more animal matter in their diet than their more arboreal counterparts. The shrew-faced ground squirrel even eats earthworms. Perhaps the most remarkable ground squirrel in appearance is the tufted ground squirrel of Borneo, a large animal with bushy tail and enormous tufts of dark hairs on the tips of its ears. Borneo also has the smallest squirrel in the world, the tiny pygmy squirrel, only 11 centimetres long, which lives in the lower layers of the forest and scampers jerkily up the tree trunks, nibbling on the bark.

Niche separation is well illustrated by the tree shrews, an interesting group of endemic Oriental mammals. They are primitive insectivorous animals and should probably be included in the same order as the shrews and hedgehogs. For many years, however, they were regarded as primitive members of Man's own order, the Primates. With our increased knowledge of their breeding behaviour this idea is losing support. The Malays have always classified these animals with the squirrels, which they resemble closely in both appearance and behaviour. Their dentition is clearly not rodent-like, but science has nevertheless adopted the Malay name *tupaia tana* (squirrel of the ground) to describe the most abundant species.

The common tree shrew has a short bushy tail and a long, pointed, sensitive nose. It is not as arboreal as its name implies and may be seen scurrying about among the forest litter or on old dead logs, probing out beetle grubs, earthworms and insects. Tree shrews travel singly or in small parties. They are jerky, nervous movers and when alerted sit up tall with head held high, scanning the forest for any sign of danger. A flick of the tail, a snort of alarm and the animal is scuttling back to its hole beneath a log.

The slightly smaller slender tree shrew has a shorter nose and longer tail. It also lives on the forest floor but exploits a different niche from the common tree shrew. It does not probe among leaf litter and dead wood but concentrates on surface insects, ants, centipedes, etc. and fallen fruits. The lesser tree shrew is smaller still, with a very much shorter nose. It is more arboreal than the other two species and feeds mainly on insects among the lower branches and saplings of the forest.

Tree shrews are territorial. They mark out their ranges by rubbing twigs with a secretion from special cutaneous glands and probably add urine. One can sometimes watch fierce noisy chases around these marked boundaries. In their breeding behaviour they are rather unusual. Adults pair and share a common nest-hole, but the female makes a separate nest to give birth to her young. The male never enters this nest and the female does so only to suckle her offspring once every two days. The young can survive on such infrequent attention because the mother's milk has a high fat and protein content. After five weeks on this rich diet the juveniles are weaned and able to leave the nest.

Life in the treetops

Most of the available sunlight is trapped in the upper storey of the forest. Consequently this is where most of the products of photosynthesis are found, the primary foods, leaves, young shoots, buds, flowers and fruit. Feeding on these and living in the upper layers of the forest places a major locomotor constraint on the arboreal animals. Only small, agile creatures can travel with ease through the maze of slim branches, leaping accurately from tree to tree. Flying birds, bats and insects are at a great advantage here but a fascinating array of squirrels, monkeys, civets, snakes, lizards and even frogs are able to live in the canopy quite independent of the forest floor.

The arboreal way of life does, however, demand certain specializations. Squirrels and civets can move at speed up vertical trunks by digging their sharp claws into the bark. Monkeys grasp supporting branches with their strong, prehensile hands and feet. So that they do not cut themselves when they curl their fingers round a slender twig, they have flat finger nails rather than claws. The palmar surfaces of their hands and feet are beautifully designed, finely-ridged pads which ensure the maximum contact and tightest grip on any support.

Monkeys are wonderful leapers, easily covering gaps of 6 metres or more before they crash into the branches of the next tree. As it leaps a leaf monkey gives a powerful kick-off with its long hind legs, and while actually airborne it rotates its long tail to maintain its balance. Gibbons, the smallest of the apes, leap in a different way. They are highly skilled exponents of a type of locomotion called brachiation. Hanging suspended by their hooked hands and long arms, they cartwheel along beneath the branches. Arm over arm, they swing through the treetops, launching themselves across wide gaps in the canopy with powerful pulls. The success of their mode of travel can be judged from the fact that gibbons move faster, more quietly and farther each day than any other forest apes or monkeys.

At the other end of the spectrum is the great, ponderous ape, the orang-utan. At 80 kilograms the male orang-utan is by far the largest truly arboreal animal in the world. Treetop life for such a heavy creature would be impossible if it did not possess remarkable hooked hands, hand-like feet, long arms and tremendous strength. These adaptations allow the orang-utan to hang tirelessly, for hours at a stretch, suspended by only three limbs while picking fruit with its spare hand. It can even hang upside down, clinging on with only one foot.

Orang-utans have developed a peculiar type of four-handed climbing by which they can move slowly through the canopy, but in spite of this they rarely exceed more than 500 metres in their daily travels. If they come to a gap between trees they are unable to leap across and all save the largest males, who find moving through the trees difficult, are unwilling to descend to the ground. Usually the animals solve this problem by rocking their tree backwards and forwards until it swings sufficiently close to its neighbour for them to grab at the branches, pull the two crowns together and scramble across.

Forest primates typically live in groups, each group occupying its own range or territory. Leaf monkey troops may contain anything from three to twenty or more animals, occupying a territory about 30 hectares in size. A mature male dominates the group, breeding with several females while the rest of the troop consists of their developing young. Female offspring stay within the group but the maturing males have two choices. They may either stay with the troop, awaiting their chance to challenge the dominant animal, or they may leave to wander alone or join up with other 'drop-outs' to form small all-male groups on the fringe of the family territories.

Throughout South-East Asia there are several

GRANDJEAN

MACKINNON

Superbly adapted for life in the treetops, the orang-utan is the only member of the great apes found in Asia. It is confined to the islands of Borneo and Sumatra, where its numbers have been much reduced by hunting to obtain young animals for zoos; but destruction of its rain-forest habitat is the most serious threat to its existence.

Left: Hoolock gibbons are the most northerly of the six gibbon species and are found in rain-forest in northern Burma and north-eastern India. Like other gibbons they are fruit-eaters and travel round their territories each day picking the ripest fruit.

MACKINNON

Silvery leaf monkeys belong more to riverine forest than to thick jungle. Like the other leaf monkey species, they are spectacular leapers, often clearing gaps of 6 metres or more.

species of leaf monkey. All eat a large proportion of leaves, as their name implies, but they show different preferences in diet. The maroon leaf monkeys of Borneo and grey and banded leaf monkeys, found in Sumatra and on the mainland, all take a great variety of young leaves, flowers and fruits. They are very shy and always stay close together, moving and feeding as a group. If they are alarmed, the forest echoes with their harsh, rattling calls. In the lowland forests of Malaya the distribution of the banded leaf monkey overlaps with that of the dusky or spectacled leaf monkey but they do not clash ecologically as the dusky leaf monkey is the more specialized feeder, eating fewer species of leaves and preferring a higher proportion of mature leaves in its diet. The dusky or spectacled leaf monkey is particularly appealing in appearance, with its blackened face and the white rings round its eyes. Adults have coats of dark grey but the babies are bright orange. (So, too, are the young of the related silvery leaf monkey, which occurs throughout the region and is a species belonging to riverine forest rather than thick tropical rain-forest.) Although dusky leaf monkeys form groups within a definite large territory, individual members often move and feed in small bands, maintaining contact with honking calls and re-congregating at night.

Gibbons live in much smaller family groups, a male, a female and their growing young. They are vigorously territorial, spending half an hour each morning calling and displaying to defend their rights or indulging in fights and chases with their neighbours over disputed boundaries. Gibbons are fruit specialists. Every day they travel quickly round their large territories feeding at many dispersed food sources, picking only the ripest fruit and leaving the rest for the next day.

Gibbons occur throughout South-East Asia but geographical separation has led to the formation of distinct species. Where two species meet there is no overlap since their ecological requirements are too similar to allow them to co-exist in the same place. The exception to this rule is the siamang or black gibbon which is twice the size of the other species. The siamang's distribution overlaps with that of the black-handed and white-handed gibbons in both Malaya and Sumatra. Because of their extra bulk siamangs need more food each day than gibbons and where the two species meet at a food source, such as a fruiting fig, the bigger siamang will chase the gibbon out of the tree. Siamangs are less mobile, however, and during their daily ranging travel only about 1 kilometre, only half the distance

The red-tailed racer is one of the larger arboreal snakes of the rain-forest, where its mainly green coloration makes it difficult to see among the leaves. It is not poisonous, but is strong and active and kills its prey – mainly rats and squirrels – by constriction.

The green *Calotes* is a common rain-forest lizard and is often seen in gardens and by roadsides, climbing actively in trees and bushes. It is normally coloured bright green but rapidly turns dark brown or black if it is captured and handled. In its sexual display the throat of the male turns bright red.

alone again to give birth and rear the infant. These females occupy relatively small home ranges but, unlike the territories of gibbon families, these are not exclusive. Males wander rather further so that their ranges may overlap with those of several potential mates. At times orang-utans show seasonal movements, ranging throughout the lowlands during the summer fruit season but retreating to the forested foothills at wetter times of the year.

Adult male orangs grow very big and achieve a quite spectacular appearance with long hair and beards, fatty crowns and gross, flapping cheek flanges. Periodically they emit a series of deep roars or 'long calls' which seem to act as a threat, keeping other males at bay, but possibly also attracting females. An old calling male will not tolerate other males nearby and will either chase them off or frighten them away with violent branch-waving displays. Since younger non-calling males seem just as successful as their elders at finding mates it is difficult to understand the significance of the 'long calls' and the rivalry and competition between adult males.

Like the African great apes, orang-utans make a new leafy nest or sleeping platform each night. They bend several strong branches inwards and weave and fold them together to make a springy bed. Often the resting animal will pile more twigs over its head. These nests remain visible for several months and are a good indication of where orangs have been active.

Orang-utans and the other Asian apes are true arboreal creatures but one species of forest monkey, the pig-tailed macaque, is equally at home on the ground or in the trees. These big baboon-like animals live together in large troops that march purposefully over the forest floor but climb high into the trees to feed. Despite their short, curly tails, they are agile jumpers and climbers. Pig-tailed macaques range over a wide area, the various troop members strung out through the forest, maintaining contact with deep grunts. They are frequently trained by the Malays to pick coconuts and become very expert, knowing just which nuts to twist and drop and which to leave.

Primates and squirrels are not the only quadrupeds to have become treetop dwellers; many civets, martens, tree shrews and rats spend much of their time in the trees. Reptile groups have several arboreal representatives: tree monitors, green *Calotes* lizards, frilled agamid lizards and most of the forest snakes. One of the commonest tree snakes is the small, delicate whip snake, which glides

covered by the gibbon. In consequence siamangs cannot afford to be so selective about their food; it is not worth their while to travel long distances to small crops. They therefore concentrate their feeding in the few most productive trees within their territory and often supplement their diet with large quantities of leaves. Gibbons feed over a much bigger area, taking the pick of the crop from the good trees when the siamangs are absent but also feeding at the smaller, more dispersed fruit trees which the siamangs do not visit.

The orang-utan is unusual for a higher primate in being an anti-social animal. Orangs are solitary creatures. Males live alone and females are usually accompanied only by dependent offspring. Occasionally they link up with males to form brief reproductive consortships but the female is left gracefully among the branches. It is usually a vivid green in colour to blend with the leafy background but pink and yellow forms also occur.

One of the most spectacular success stories of an animal adopting the arboreal way of life is that of an amphibian, the tree frog. To enable them to leap from branch to branch and tree to tree, these frogs have suckered toes, which adhere firmly to the leaf surfaces on which they land. They have large eyes, an adaptation to nocturnal living, and their curious tonking calls are among the most characteristic sounds of the jungle night. The real secret of their success, however, is their independence of standing water for egg-laying and tadpole development. They lay their eggs in a wet frothy mass attached to a large leaf in the damper parts of the canopy. The froth protects the eggs from evaporation and the growing tadpoles remain within this sheath until heavy rain comes. Then they are washed down into crevices in the bark where they complete their metamorphosis and emerge as tiny tree frogs, fully equipped for arboreal life.

Forest gliders and bats

The advantages of being able to get from one tree to the next without having to descend to the ground and climb up again are manifold. Any animal with this ability avoids coming into contact with large predators such as the leopard and tiger which roam the forest floor. It also saves time and energy and the animal is, therefore, able to travel over a wider area each day in search of food. The ability to glide extends its range even further.

Gliding has been developed independently among six different animal groups that exist in the Oriental rain-forest. The most common of these is the flying lizard, *Draco*. There are several species of *Draco*, all living on tree-trunks and all feeding on ants, but occurring at different heights in the forest. Males are territorial and erect colourful flaps of skin beneath the throat to warn off trespassers and attract females. More extraordinary, however, is their ability to erect lateral flaps of skin along their body. The flaps of skin are supported by six bony rays, extensions of the ribs, and can be extended like wings, enabling the lizard to leap from its support and glide gently to another. Like other agamids, the female *Draco* must descend to the ground to lay her eggs in the soil. In this respect the flying gecko has an advantage; it cements its eggs to the sides of a tree-trunk and is thus quite independent of the forest floor throughout its whole life

POLUNIN/NHPA

MACKINNON

The Asian rain-forest supports a variety of animals which, without really flying, can glide or parachute through the air. Flying lizards (above) have a membrane of skin on each side, supported by prolongations of the ribs. This is spread out as they glide from one tree or branch to another, and folded along the sides when the lizard is running or resting. Flying lemurs (right) are even more spectacular gliders. A wide flap of skin between their outstretched limbs and around the tail, enables them to cover as much as 100 metres in one glide. Flying snakes (above, right) are perhaps the most unexpected of the gliding animals. The paradise tree snake has no extra 'flap' of skin, but becomes airborne by flattening its whole body and hollowing it underneath.

Top: Tree frogs of the genus *Rhacophorus* are wholly arboreal; the ends of their toes are expanded, forming sucking discs with which they cling to twigs and leaves. Even their eggs are laid above the ground, often in trees overhanging water. Protected from evaporation by a mass of frothy albumen, the eggs remain attached to a large leaf until heavy rain washes them into crevices in the bark or down to the water below.

LIM BOO LIAT

MACKINNON

cycle. Like the flying lizard, the flying gecko raises lateral skin extensions to form a gliding membrane; when not in use they are folded under its belly like a waistcoat. The flattened shape is doubly useful. It lends the body extra buoyancy during a glide and helps render the gecko almost invisible when it flattens itself against a tree-trunk.

As can be seen from the flying lizards, an animal's ability to glide depends on increasing its surface area in contact with the air. Gliding animals have different ways of achieving this. The rare flying frogs have greatly expanded webbing on their feet which allows them to parachute gently downwards from tree to tree. The golden flying snake slithers quickly along a branch and, without checking its speed, flings itself across a gap. As it takes off the snake flattens its body so that the ventral surface is hollowed, and in mid-air it assumes a tight S-shape.

The gliding performances of the flying lemur and flying squirrels, of which there are many genera both large and small, are even more spectacular. From a take-off point high on a tall tree these animals can easily cover 100 metres in a single glide. Both groups have a wide flap of skin, a flight membrane or patagium, which stretches between their outstretched limbs, and in the lemur also encompasses the tail. The flying lemur gains additional span because of the extreme length of its fingers and toes. Flying squirrels increase their surface area by means of a stiff cartilage rod which can be extended beyond the hand, drawing the patagium with it. This rod and the squirrel's tail may be employed in steering. These gliding mechanisms, however, render the animals clumsy climbers, very vulnerable to fast-flying eagles and hawks, and it is probably for this reason that both flying lemur and flying squirrels are nocturnal.

Squirrels hide in tree holes during the day, but the flying lemur remains in the open, clinging onto trunks or branches. Its mottled, almost green coat blends perfectly with its background, providing excellent camouflage. The lack of a permanent den or nest means that right from birth the infants must be carried everywhere. The single young spends most of its early life clinging to the mother's fur and when she rests, hanging beneath a horizontal bough, her parachute provides it with a natural hammock.

The only mammal group to have developed true, powered flight is the bats and these are well represented in the rain-forest. Like other members of their family, most forest bats are insect feeders

but several feed on nectar, landing on the flower and licking out the nectar with their long, thin tongues. Bat-pollinated plants bear large, plumose flowers, usually white and purple in colour so that they will show up at night, and include such cultivated species as bananas and durians.

The biggest and most impressive bats of all are the giant fruit bats, or flying foxes, which may have a wingspan of up to 1·5 metres. They are gregarious animals, sharing common daytime roosts where tens of thousands hang head down from adjacent trees, fanning themselves with leathery membranous wings and uttering shrill squeaks. Fruit bats are migratory. The colony may use the same tree roost for months, then suddenly vanish not to return until several months later. Roosts are often located on the coast or beside a river, not so much for reasons of safety but rather because the bats use the coastline or waterway to find their route home again after covering many kilometres each night in search of fruit and young leaves.

Life on the forest floor

Little light reaches the forest floor and few primary food-producing plants are accessible to ground-living animals. On the other hand this bottom stratum of the forest receives all the debris that falls from above: corpses, animal faeces, dropped fruits, fallen seeds, leaves, broken branches and even whole fallen trees. These combine to form a very rich, decaying forest litter, which provides for an exotic array of invertebrate and vertebrate fauna.

The forest floor is insulated from climatic extremes by the thick vegetation cover above. Temperatures vary by as little as 5°C throughout the year. Humidity, too, is constant, 100 per cent at night and rarely below 95 per cent in the middle of the day. This high humidity enables a number of animals which would normally be regarded as aquatic to live terrestrially in the damp forest. They include a multitude of blood-sucking leeches. Looping over the fallen leaves, they wave their bodies in the air to sense the approach of any animal large enough to provide them with a meal. When one comes within range, they attach themselves to it by means of a suckered mouthpiece and are carried around with the host until their bodies are distended with blood, when they release their grip and fall back to the forest floor.

More decorative, and less common, are the terrestrial Turbellaria or flatworms. Colourfully

SOEPADMO

HARDING

Rhinoceros beetles belong to the Scarab family and some, like this one, are large and impressive insects. They have two forwardly projecting horns, one on the thorax and one on the head; up-and-down movements of the head cause the horns to operate rather like a pair of forceps. One species is a serious pest of coconut palms.

Top: The small, long-tongued fruit bats have elongated muzzles and long tongues with brush-like tips – an adaptation for feeding on the nectar of flowers. Like other nectar feeders they convey pollen from flower to flower and some plants, including *Oroxylon indicum* shown here, are dependent on bats for pollination.

Left: When the flying lemur is at rest, its web of skin forms a natural hammock for its young.

45

striped, with sickle-shaped heads, these creatures flow slimily across the damp vegetation. They feed on organic material by means of an eversible mouth situated on the underside, halfway down the body. These curious primitive worms can reproduce merely by splitting off buds from the tail.

Far from water, crabs scuttle about the forest floor at night and a great variety of frogs live in holes under tree roots or in clumps of wild ginger. After a heavy rainstorm puddles of standing water may remain for several days. Then the forest is filled by a deafening croaking as hundreds of frogs congregate to mate and spawn. Macaques, hawks and snakes arrive to take their toll but many survive. Usually the water evaporates or drains away before the tadpoles are fully grown but sometimes the wet weather continues for long enough for a new generation of tiny frogs to develop.

Dead wood is a habitat in itself. It has its own associated fauna and flora, for which it provides food and shelter and which in turn help to break down the woody material and recycle minerals and nutrients into the ecosystem. The organisms of the detritus food chain are called decomposers and include armies of termites, pill millipedes, fierce-jawed stag beetles, rhinoceros beetles and the beautiful, metallic green longhorn beetles that are used for ornamental jewellery. Fungi, too, play their part, their spreading hyphae insinuating themselves among the woody tissue. Luminous toadstools and delicate cup fungi spring up on rotting logs to be devoured by ants and by hordes of black and yellow beetles. The extraordinary basket fungus exudes a foul smell of decaying meat, attracting blowflies, *Charaxes* butterflies and long-headed grasshoppers which feast on the slimy cap and help cross spore-types with other basket fungi they visit.

Even more curious are the fiddle beetles, members of the carnivorous family Carabidae. They were once thought to be very rare and, in the middle of the last century, the Paris Museum paid 1,000 francs for a single specimen. Subsequent investigation into their life history has shown that they are quite common. The nocturnal adults are found resting on bracket fungi which grow on rotting wood; the beetle's flattened shape is well suited to life among the plate-like fungi. The female bores a hole in the fungus where she deposits her egg and the developing larva feeds on tiny insects which enter its chamber.

Other carabids include the gaudy tiger beetles which can be seen scuttling along fallen trunks or

MACKINNON

ROSS

A column of worker termites forage for food on the forest floor, guarded by big-jawed 'soldiers'. Most termites are nocturnal but some species, usually black in colour, are active by day.

Top: The beautiful basket fungus develops a net-like veil, called the indusium, which hangs in folds from the small cap. Basket fungi have a foul smell, like that of carrion, which attracts flies. In this way the spores of the fungus are dispersed by a mechanism almost exactly like the insect pollination of flowers.

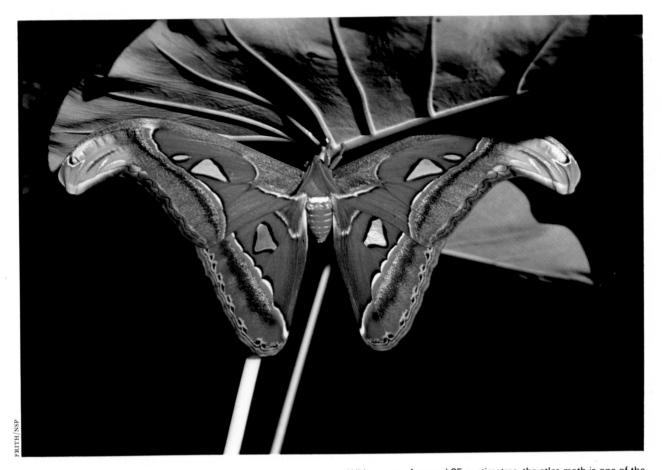

With a span of around 25 centimetres, the atlas moth is one of the largest moths in the world. In the forested region of tropical Asia it is not uncommon and it sometimes flies into lighted houses. Its huge, pale green caterpillars feed on a variety of trees and bushes and are often found in gardens, where a dozen of them may completely defoliate a small shrub.

Numerous kinds of spiders live among the leaves of trees and bushes in tropical forest. Some, like this sun spider, have long, thorn-like projections from their bodies. These may give them some protection against birds which swallow their prey whole.

ROSS

MACKINNON

WARD/NSP

Camouflage in insects and other small animals is usually a purely protective device, serving to conceal the insect from its enemies If the insect is itself a predator, the camouflage may serve the double purpose of deceiving both its enemies and its prey. The orchid mantis (top) is coloured pink or white in its subadult stages and has a segment on the hind and middle legs expanded to resemble the petals of a flower. The dung spider (above, left) is coloured to resemble the white excrement of a bird and adds to

the effect by sitting on an irregular white web of silk. Some insects feed on fresh bird-droppings, as well as on the nectar of flowers, so these predators are both protected by their appearance and assisted in ambushing their prey. The bush cricket or katydid's cryptic coloration (above, right) is purely protective. Its wings are shaped, coloured and veined almost exactly like leaves and it even has markings which look like those produced on real leaves by fungi.

foraging in the forest clearings. They are voracious predators, rushing to seize any small, slow-moving creatures which amble into sight. The larvae of the tiger beetle occupy burrows in the ground and lie in wait, their gaping jaws level with the surface, to snap on any creature that passes overhead. The unfortunate prey is pulled down into the burrow to be devoured at leisure.

Numerous other predators lie in ambush for unwary insects. Armies of ants hunt and scavenge among the leaf litter. Colourful sun spiders and fat orb spiders spin their delicate webs to imprison flying insects. Crab spiders merge into the flowers they rest on, snapping on bees and butterflies that come to feed at the nectar. The incredible orchid mantis is its own flower, with pink or white 'petals', and attracts food straight into its waiting arms. A similar trick of disguise is employed by the dung spider. This little, shiny black and white spider sits with outstretched legs on a white silk mat attached to a leaf, looking just like a bird-dropping. The disguise has a dual purpose: spider-hunting wasps take no notice while flies attracted to the 'dung' are caught and eaten.

As a protection against the numerous types of forest predators, many invertebrates have de-veloped hard shells, spines, jaws or foul chemicals. Excellent examples of such defences are found among the millipedes. The familiar pill millipedes that crawl slowly about the forest will, at the slightest disturbance, snap shut to form a perfect sphere of hard impenetrable cuticle. Some of the larger species may form spheres as big as a ping pong ball. They take a long time to open again, and do so very cautiously, by which time any civet or bird will have given up and gone. Still larger red and black millipedes, some exceeding 20 centimetres in length, glide smoothly over tree-trunks. These also curl tight when disturbed, coiling up into a flat disc, like a fossil ammonite, but leaving their myriad legs exposed. Any further attempt to molest the creature will be countered with a flood of orange-staining oil of unpleasant odour and stinging taste.

Many of the forest insects employ disguises to protect themselves from predators. Crickets, leaf insects and butterflies at rest are camouflaged to look like the surrounding foliage. They have developed intricate patterns of veins, disease spots and even leaf-cutter scars to make their disguise more complete. Moths and stick insects when stationary resemble twigs and bark so closely in

Stick insects rely for protection on concealment. If they are dis-covered, they may resort to a startling display, standing high on their legs and flashing brightly coloured wings, in an attempt to frighten the predator that has penetrated their disguise.

Some of the millipedes of the tropical forest are 20 centimetres or more in length, giants in comparison with the small species familiar in temperate regions. When molested or handled they coil in a spiral and often exude a corrosive fluid from pores along the sides of their body – a protective device to deter predators.

colour and form that they are indistinguishable from them. Such camouflages are not confined to insects – they are found throughout the animal kingdom. The horned frog, for example, is cryptically coloured so that it blends perfectly with the leaf litter on which it sits and, among birds, the nocturnal frogmouth perches by day in such a way that it looks like a branch of its tree roost.

Mimicry is another protective mechanism employed by several insect groups; the trick here lies in resembling another animal so closely that a would-be predator cannot differentiate between the two. Many poisonous, distasteful or well-armed insects are conspicuously coloured, advertising themselves to the full. Black and yellow stripes, such as those of wasps, are a good indication of distastefulness. After a few attempts predators learn to avoid animals displaying the characteristic patterns. Thus any animal who 'cheats' and adopts the same warning colours will be left alone, even if it is in fact perfectly palatable. In some cases the 'mimics' duplicate the patterns of their unattractive 'models' so exactly that they fool naturalist collectors let alone the predacious toads, lizards, birds, ants and spiders that take a heavy daily toll of forest insect life.

Another mechanism, the 'startle' response, is often used as a second line of defence by camouflaged insects. Species such as crickets, mantids and stick insects, which resemble twigs and leaves and are dull and inconspicuous at rest, will flash brightly coloured underwings if they are disturbed by a potential predator. The sudden transformation from inconspicuous immobility to active colour startles the predator and gives the insect time to escape. Drab moths often open their wings to reveal huge owl-like 'eyes' which frighten off vertebrate attackers. Small eyespots on the hindwings of cryptic butterflies serve a rather different function. They deflect the predator's attack away from the insect's vulnerable body. In the blue butterflies filamentous false 'antennae' reinforce the impression of a head at the wrong end of the body.

The markings of the common jungle glory butterfly, which feeds from rotting fruit on the forest floor, are a good example of protective coloration. As it rests on the ground the butterfly resembles a dead leaf but closer inspection reveals small eyespots near the tail. If disturbed it flashes open its wings to reveal an upper surface of dazzling, metallic blue. The insect flies with a slow wingbeat and is clearly visible while the wings are

MACKINNON

Mousedeer or chevrotains are among the smallest of all hoofed mammals, and have very slender legs and tiny feet supporting a rather rounded body. They are not true deer and have neither horns nor antlers, though the males have enlarged canine teeth in the upper jaw.

open but when it folds its wings and resettles it is easily lost by a pursuer.

Fallen fruit provides a food supply for a host of creatures on the forest floor. Cautious tortoises are attracted to the debris. Adults are well protected by their hard box-like shells but the young tortoises are softer and easier for tigers and other predators to open. To compensate for this their shells are covered with sharp spines, which must make them a very prickly mouthful indeed. As the animal grows the points are gradually worn away until the shell of the adult bears only a few tell-tale bumps.

Several forest mammals feed on the fallen fruit. Tiny mousedeer, no bigger than a hare, are quite common. Active by night and day they travel singly or in pairs. They are not true deer though they are thought to be similar to the deer's evolutionary ancestors. Instead of antlers they have developed long, curved canines for defence. The muntjac, or barking deer, bridges the gap between the mousedeer and the true deer; it has both long canines and short, pointed antlers. Muntjac stags show territorial behaviour. Their fierce barking keeps other males away but attracts the does to them. The Bornean natives have learned to imitate this call on a leaf whistle and shoot the doe or challenging stag as it approaches.

Probably the most successful of the mammals that feed on the fallen fruits are the wild pigs. The common wild boar is a ubiquitous forest mammal, ranging from north-west Europe to south-east Asia. Tropical boars are smaller and less hairy than their European counterparts but show similar herd structure and diet, rooting about for worms and fallen seeds and fruit. The bearded pig occurs only in the rain-forest. It is larger than the common boar and has a bristly beard. In the Bornean forest the bearded pig is the most abundant mammal (measured in terms of biomass). As we have mentioned earlier, many of the jungle trees are dipterocarps, which fruit seasonally in any locality. For much of the year there are few seeds available in any one place, then, suddenly, in the fruiting season there are too many for the squirrels to cope with and most of the crop falls to the ground to benefit the pigs. The bearded pig takes advantage of the fact that the same trees in different valleys fruit at different times of the year and the herds follow long-established migratory routes that bring them to the various feeding areas just as the crops ripen and fall. In good years bands of thousands of migrating pigs root through the forest, swimming across wide rivers, and even sea coves, in their quest for fresh supplies of food.

Both species of pig produce large litters of striped piglets, born in great piles of vegetation bitten off and collected by the mother sow. Similar structures are sometimes made by wild pigs for shelter from heavy rainstorms. These 'nests' are usually alive with the unpleasant ectoparasites of the previous occupant and human travellers through the jungle are well advised to keep away.

Before leaving the forest floor it is worth pointing out that within such a habitat there are other, smaller habitats. Physical conditions within a rotting log, under bark, in a hole in the ground, among the flowers and foliage of shrubs may be very different from those prevailing generally on the forest floor, and these microhabitats can support their own characteristic fauna. Corpses and dung are two which provide specialized niches. Apart from attracting the scavengers, beetles, flies and ants that tackle these 'delicacies' they provide food and shelter for developing maggots and the larvae of dung beetles.

Some animals, such as nest-builders or burrowers, can be said to furnish their own microhabitat. The mud-dauber wasp goes a step further by constructing and provisioning a chamber of

Young spiny tortoises have sharp spines on all the marginal plates of their shell. These may deter predators, and they certainly help to conceal the tortoise, giving it an irregular outline when it is motionless against a background of dead leaves. In the adult the spines are blunt, and are only developed in front and behind.

MACKINNON

51

RAO/NHPA

The wild boar is widely distributed in Asia, Europe and North Africa and is the ancestor of the domestic pig. It is a hardy animal, strong and active, quite omnivorous and able to maintain itself well even when it is hunted – as it is almost everywhere it occurs. The young, as many as twelve in a litter, are striped.

Right: The Asian elephant has been domesticated for centuries and is remarkably docile and intelligent. Elephants are particularly useful for handling heavy timber and will pick up large logs and stack them in a most orderly way. A working elephant is always controlled by a rider, known in India as a *mahout*.

clay in which its single egg is completely sealed off from the outside environment. Though these chambers are usually attached to rocks or tree-trunks close to the ground the wasp must exploit all layers of the forest to collect the necessary building materials. A drop of water is carried from a puddle or stream and mixed with dry earth to make the mud pellets with which the chamber is built. The wasp then hunts in the canopy, catching and paralysing spiders which provide a store of fresh food for the developing larva.

Elephants, rhinoceros and tapirs

Asia's largest land animal, the Indian elephant, has a wide distribution throughout the region. Basically it is a forest dweller but it can exist anywhere providing there is rich vegetation and an adequate water supply. Today elephants are found in Ceylon and India, through Nepal to southern China and southwards through Indo-China and Malaya to the greater islands of the Sunda Shelf. Once elephants roamed across Java but the species is now extinct there. During the Pleistocene pygmy forms even reached Celebes. In many parts of their range elephants have been domesticated and are now used for carrying heavy loads or for human transport.

Elephants are unusual in that they have a matriarchal society with bulls living peripheral to the close-knit mother-family groups. In wide open feeding areas the huge beasts congregate in herds but in the denser forest they travel in much smaller bands, a mother and her calves or a solitary old bull. Elephants are browsers and as they crash through the jungle they tear down rotans, vines, bamboo and small palms on which they feed. They will also take fruit, the roots of ferns and wild ginger, and, where it is available, riverside or swamp grass. Elephants are extremely destructive feeders, tearing down branches and uprooting young shrubs. Usually they travel slowly over large areas so that damage is scattered but where they occur at high densities their destructive habits seriously affect the growth pattern of the forest and encourage the spread of vine-tangled secondary jungle.

Elephants cover great distances, following ancient traditional paths that link their favourite feeding areas, mud wallows, drinking pools and mineral licks. In Sumatra, where much of the soil is leached of salts, they excavate deep caves into mineral-rich rock to obtain vital sodium and potassium. The great bulls use their tusks to gouge

out lumps of rock from the roof but the tuskless cows and calves must use their feet to dislodge the rich dust from the sides of the cavern. Other animals benefit from these excavations: sambar deer, muntjac, porcupines, squirrels, monkeys and even orang-utans all visit the caves to eat the mineral-rich soil.

Three species of rhinoceros are found in tropical Asia and were all once widespread. Today they are extremely rare and localized, hunted out of existence for their much-prized horns, which the natives believe to have powerful medicinal and aphrodisiac properties. The bulky Indian rhinoceros is now confined to small areas of riverine grassland in Assam and Nepal where only a few hundred survive. The Javanese rhinoceros, another one-horned species, is perhaps the rarest of all large mammals. It is now probably extinct on the mainland and Sumatra but can still be found in the Udjong Kulon peninsula on the westernmost tip of Java. Here, in the low-lying swampy forest, a mere two hundred survive.

The Sumatran rhinoceros has two horns. It is the smallest of all living rhinoceros and also the hairiest, with a coarse, russet coat. It is an inhabitant of hill forest and its range includes Burma, Thailand and Malaya, as well as the islands of Borneo and Sumatra. Today these animals can only be found in scattered, small populations and they are perhaps the most threatened species of all. The rhinoceros wander singly or in pairs over wide areas, kicking down saplings to browse on their leaves. They are fond of wallowing in mud to cool off and to discourage skin parasites and regularly revisit the same wallows. This habit has made them extremely vulnerable to hunters and poachers. Sumatran natives hoist heavy logs over the jungle paths and attach bamboo spears to these weights. When the bulky rhinoceros knocks against a trip trigger the boom is released and the animal is skewered to death. On the Asian mainland they are captured in concealed pit fall traps.

Related to the rhinoceros is the curious Malay tapir. The fact that tapirs are found only in South-East Asia and South America led some early zoologists to postulate a land bridge connecting the two continents, in spite of the fact that they share no other common mammal. More recent discoveries of fossil tapirs in Europe and North America have shown that this primitive group once enjoyed a world-wide distribution, of which the two surviving species are merely relicts. Though

Sumatran rhinoceros

Tapirs now exist only in South-East Asia and tropical America though in former geological times they had a world-wide distribution. Malayan tapirs prefer swampy, low-lying forest, gathering roots and palm shoots with their short trunks, and visiting salt-licks for minerals. They differ from American tapirs in being conspicuously particoloured, black and white, though the young of all tapirs are striped white on a dark ground.

Three of the world's existing species of rhinoceros live in tropical Asia. At one time they were widespread, but their range has been so reduced by hunting that their natural distribution is no longer apparent.

The Great Indian rhinoceros has always lived in north and north-eastern India, where it is now protected in three widely separated areas. Over 2 metres high at the shoulder and over 4 metres long, it weighs 2 tonnes. The slightly smaller Javan rhinoceros is a lowland forest species that once ranged from north-eastern India to Java, Malaya and Sumatra but is now known only in the Udjong Reserve in Java. It is slightly over $1\frac{1}{2}$ metres high, $3\frac{1}{2}$ metres long and weighs 1 tonne – just half the weight of the Great Indian species. The smallest of the three, the Sumatran rhinoceros, once had a similar distribution in hilly country (including Borneo but excluding Java) but it is now probably the most threatened species of them all. Only just over 1 metre high, and around $2\frac{1}{2}$ metres long, it weighs 450 kilograms.

Great Indian rhinoceros   Javan rhinoceros

now extinct on Java, the Malay tapir is still fairly common in Sumatra, the Malay peninsula and the southern parts of Indo-China and Thailand.

The Malay tapir is a shy, nocturnal animal. Like many other creatures of the night it has striking black and white markings which probably serve a social function as a recognition signal. Although it is a forest dweller it prefers low-lying swampy areas, and is at least partly aquatic in habit. Tapirs feed on roots and palm shoots which they gather with their curious short trunk. They, too, frequent salt-licks to obtain necessary minerals. Newly born young have striped and spotted brown coats which afford them camouflage among the dappled jungle vegetation; in time they moult and adopt the bold piebald markings of the adult pelage.

The carnivorous mammals

The carnivorous mammals include several large and well-known types but in actual numbers they are rare in the forest. As the last step in the food chain they must, to survive, be much less common than their herbivorous prey.

The hunting, stalking members of the cat family are well represented in the Oriental region, with the tiger, leopard and a number of smaller cats. Although it can be found in forest, the tiger is typically an inhabitant of more open terrain. The leopard, however, is an accomplished forest preda-tor. Most of the leopards living in primary jungle are of the dark or melanic form and the black rosettes that form the leopard's characteristic spots can only be seen at close quarters. They are solitary animals and hunt deer and wild pigs at night, sometimes springing onto them from an overhanging branch. When they have eaten their fill, they store the remains of their kill in the crotch of a small tree to protect it from other ground predators and scavengers.

Although tiger and leopard reached Java, neither are known in Borneo where the clouded leopard is the largest cat. This magnificent animal, almost a metre in length with a long tail, is semi-arboreal and preys on monkeys and wild pigs. It does not occur west of Sumatra but the smaller fishing cat, which lives on river banks, is much more wide-spread and still ranges through parts of India and the marshlands of Ceylon. It does not enter the water to fish, but crouches on a rock or an over-hanging bank, scooping up the fish with its webbed forefeet. Elsewhere are found the fierce little leopard cat, slightly smaller than a domestic cat,

The leopard cat is about the same size as a domestic cat, though its legs are rather longer. It lives in a remarkably wide variety of climates, from the severe cold of Tibet and Siberia to the hot rain-forest of Malaysia and Indonesia.

The black panther is no more than a melanistic variety of the more widespread spotted panther or leopard; it has the same spotted pattern, but its spots are hard to see against the dark background of its fur. Kittens of both forms may appear in the same litter, though the black variety is predominant in the humid rain-forest of tropical Asia.

and the medium-sized cats: the bay cat, flat-headed cat, marbled cat and the larger golden cat. All are nocturnal, solitary hunters, preying on slightly different ranges of food, vertebrate and invertebrate.

The red-coated dhole or wild dog occurs throughout central Asia, Indo-Malaya, Sumatra and Java. Hunting in small packs, they chase down deer and pigs in the forest. A single pack hunts over a very wide area and with increased deforestation throughout their range these animals are becoming very rare.

Bears reached tropical Asia from the Palaearctic Region to the north. Two species dwell in the rainforest, the sloth bear of India and Ceylon and the sun bear, which occurs from Thailand and Burma south to the Sunda Islands, but is absent from Java. Although they belong to the Carnivora both these bears are, in fact, omnivorous. They eat fruit and grubs, raid birds' nests for eggs and young, and are specialists in the art of opening up the nests of bees and termites. Sun bears are small and skilful tree climbers. They prefer the nests of the stingless meliponid bees which are built deep within a tree-trunk. With its powerful jaws and clawed feet the sun bear rips open the bole and uses its long, pointed tongue to clear the nest of its sweet contents. The heavier sloth bear is expert at breaching termite mounds and it has an adaptation specifically for this purpose. It lacks the two central, upper incisor teeth and can therefore feed by pushing its nose, with nostrils tightly closed, into the breach and noisily sucking the termites through the gap in its teeth.

Sloth bears are also unusual among bears in that the cubs often ride on the mother's back. Both sloth and sun bears are dangerous, bad-tempered animals, especially when accompanied by young. Because of their poor eyesight they often blunder too close to men working in the forest. Realizing their predicament, they panic and lash out with their sharp claws and vicious jaws, inflicting serious injuries.

The mustelids (stoat family) are small carnivores found throughout the world. In the tropical rainforest they are represented by a golden-red weasel and the bigger arboreal yellow-necked marten. The marten has a mixed diet of honey, bees, insect grubs and crustaceans but it also takes larger prey such as squirrels and will even tackle a mousedeer. The teledu, another member of the stoat family, has specialized in feeding on earthworms and has developed a blunt, pig-like snout for burrowing

Fishing cat

Clouded leopard

Golden cat

after its prey. The teledu is black with a bold white stripe along each side. It is similar to the American skunk in appearance and, like the skunk, has a proverbially bad odour.

The last group of carnivores to be discussed were probably the first to reach tropical Asia. These are the Viverridae, the family of mongooses and civets. Typically, mongooses are inhabitants of the open bush but the collared and short-tailed mongooses of the Greater Sunda Islands dwell in the rain-forest. They are terrestrial animals, active mostly by day, and feed on cockroaches, small vertebrates, eggs and even snakes. They occupy dens dug in banksides, under trees or in rock crevices.

The civets are rather primitive carnivores, feeding mainly on fruits and insects rather than flesh. In the absence of other carnivores they have been able to diversify and specialize in several different niches. Two, the common civet or tangalung and the banded palm civet, are terrestrial. The tangalung, active mainly at night, is an unspecialized form with a generalized dentition able to cope with a catholic diet of fruit, insects, earthworms, small vertebrates and carrion. The banded palm civet is also nocturnal but it feeds almost exclusively on crickets and worms on the forest floor.

Seven species of wild cat, ranging in size from the clouded leopard (195 cm including tail) to the little leopard cat (90 cm) live in the rain-forest. Most take small birds and animals as prey, though the clouded leopard can kill deer, goats, even wild pigs; the fishing cat has specially adapted forefeet, webbed, with projecting claws, which enable it to catch fish from the river bank.

Found only in southern India and Ceylon, the sloth bear is unique among bears in being adapted to hunt and eat insects, though it does take fruit and other food as well. The nostrils can be closed by muscular action and the central upper incisor teeth are missing. The bear tears open termites' nests with its large front claws, then sucks the insects up through the gap in its teeth.

Marbled cat

Bay cat

Flat-headed cat

Leopard cat

Apart from the otter civet, an aquatic fish-eating species, the other civets are all mainly arboreal. The long, slinky linsang has sharp, pointed teeth, adapted like a cat's for preying on birds. The common palm civet has much broader teeth suited to the high proportion of fruit in its diet. The largest civet of all is the binturong, a grizzly, long-whiskered animal with a thick, strong prehensile tail. Although primarily nocturnal, it may be seen in the middle of the day feasting in a fig tree. The binturong feeds almost exclusively on fruit though no doubt it will supplement its diet with eggs and fledglings if it comes across an accessible nest.

Forest birds

Of all the animals living within the tropical rainforest, birds are the most often seen and heard. Like the other groups, birds can be found in every stratum of the forest, occupying every niche from fruit-eater and insect-eater to predator. It is not possible to discuss here all the different species that occur in the forest. Instead, descriptions are limited to the most important and most typical birds of this habitat.

The members of the hornbill family are large birds with powerful beaks. They feed in the canopy, feasting on fruit and insects, crickets and termites plucked from among the foliage. Hornbills toss their food several times with their beak until it is sufficiently well crushed to be flicked into the gullet and swallowed.

Some of the smaller species, such as the pied and black hornbills, rear several young together but the bigger birds, including the rhinoceros, great Indian

The banded palm civet is one of the rarer members of the civet cat family. Most of the civets climb among the trees but this one is believed to live mainly on the ground, feeding almost exclusively on crickets and worms.

Right: The great Indian hornbill is one of the larger of the hornbill family, and has the characteristic casque on the top of its beak. Though the number of young produced varies with each species, their nesting behaviour is the same. The female enters a hole in a tree and partly walls up the entrance, so that she cannot get out. She is then fed by the male for a period of two months or more. When the young have hatched, she cuts her way out and reseals the entrance, leaving only a small feeding hole. As each fledgling leaves the nest, the entrance is resealed, until the last young bird has emerged.

The coppersmith barbet is so called because its repetitive call sounds like metal being beaten. Barbets resemble woodpeckers in their nesting habits, for they excavate holes in trees with their powerful bills. Unlike woodpeckers, they are fruit eaters.

GRANDJEAN

entrance is resealed until the last fledgling emerges.

Hornbills get their name from the great hollow casque above the beak, which gives adornment as well as helping to resonate the bird's harsh calls. In the rhinoceros hornbill the casque is curled into an absurd cone, just like a rhinoceros horn. The casque of the helmeted hornbill is fronted with a thin layer of 'ivory'. This *ho-ting*, as it is called by the Chinese, was a highly prized commodity used for delicate carvings and intricately decorated snuff bottles. The helmeted hornbill has become quite rare and a new demand by western collectors for Chinese *ho-ting* snuff bottles has resurrected the old craft and further endangered the species. It is not surprising that such an important bird is well-established in native folklore. Pealing out across the dense jungle, the call of the helmeted hornbill is very disturbing. The bird emits single clear hoots which gradually increase in frequency until the call speeds up into a maniacal laughing climax. The Malays call the helmeted hornbill *butong mentua* (the mother-in-law bird), likening the call to the sound of an axe wielded by a young man chopping through the legs of his mother-in-law's stilted house. The dramatic ending represents his evil chuckles as his efforts begin to take effect.

One of the best opportunities to see the fruit-eating birds of the rain-forest comes when a large fig tree is covered in ripening fruit. Solitary imperial pigeons mingle among the hornbills, monkeys, squirrels and civets. Pink-necked green pigeons arrive dramatically in great flocks each morning and evening and weave noisily to and fro, packing their swollen crops. Emerald doves feed on the ground, pecking over the fallen fruit. Pretty little Malay or hanging lorikeets, with striking green plumage and red caps, arrive to feed on the sweet figs. They are delicate feeders and can sip nectar from flowers as well as eat fruit. At night they are an amazing sight as they hang in flocks upside down, their feet firmly hooked over small twigs and their bodies dangling below.

A small blue and black bird can often be distinguished feeding among the pigeons. It plucks one whole fig at a time and carries it to a perch to eat. This is the fairy bluebird, the commonest member of a strictly Oriental family. Fairy bluebirds may feed in flocks but are often found singly or in pairs. They are active little birds and will take insects, especially flying termites, as well as fruit. Equally colourful, the barbets also flit through the treetops, feeding primarily on fruit but also taking a few

and helmeted hornbills, produce their young singly. All hornbills, however, have the same remarkable nesting behaviour. A pair will select a tree hole or cavity and begin to seal up the entrance with mud. When the female is ready to lay, she squeezes through the crack into the nest-hole and spends two or three days plastering the gap with her droppings. The male, too, continues to seal the entrance until there is only a very small gap through which he can feed his imprisoned mate. While the female lays and incubates the clutch her partner brings her food. When the young have hatched and can take food direct from either parent, the female cuts her way out of the nest, reseals the aperture and joins her mate in collecting provisions for the brood. The developing young fledge at different times but as each one leaves the nest the insects. The forest species are all bright green and can only be distinguished by the patterns of coloured patches, red, yellow and black that adorn the head. Like woodpeckers, the barbets have two toes pointed forwards and two backwards to enable them to maintain their grip on rough bark while they excavate nest-holes in tree-trunks.

Hole nesters *par excellence*, the woodpeckers themselves are strictly insectivorous. They dislodge and expose grubs from under bark or tree holes and fork them out with their stiletto tongues. Gripping with their curious toes and resting on their tails, woodpeckers are able to support themselves on almost vertical surfaces while they drill with their powerful, chisel-like bills. Different species of woodpecker occur in the various forest levels and each specializes on a different prey. By their hunting and boring, woodpeckers expose rotting trees to further attack by insects and fungi and thus play an important ecological role in speeding up the rate at which organic material is recycled.

Other insect feeders of the canopy include broadbills and drongos. These birds hunt actively. When one spots an airborne insect it launches itself from its perch, catches its prey with an audible snap and carries the prize off to another feeding perch. The racket-tail drongo is so named because of its two long, elegant tail feathers which are bare-ribbed up to the flagged tips. In Malay it is called *hamba kera* (slave of the monkeys) from its habit of accompanying bands of monkeys through the tree-tops, catching the insects disturbed by their vigorous activity.

It is probably a similar advantage that gives rise to the phenomenon of mixed feeding flocks. In the hill forests of Ceylon a single flock might comprise a collection of grey tits, white eyes, bulbuls, babblers, nuthatches, shrikes, flycatchers and even a mammal, the dusky striped jungle squirrel. As it travels through the canopy each bird will disturb insects. Since these will fly away from it at unpredictable moments, the bird has little chance of catching them itself. Another adjacent bird, however, has a much better chance. Thus in the long run each bird in a flock does better than if it foraged alone. The mixture of species helps to reduce competition for any specific food type.

Flocking behaviour has another advantage – members of a flock are less likely to be caught by predatory hawks. With several birds on the alert there are more chances that the danger will be spotted and the alarm raised. Moreover in the

ROSS

FOGDEN

Pittas live on the ground in thick forest, obtaining their insect food by scratching about in the leaf litter. Almost all of them are beautiful birds with brilliantly variegated plumage, like this banded pitta from Malaysia and Indonesia.

Top: Parakeets are medium-sized parrots, with long tails and mainly green plumage. Several species are common in Asia and are often seen flying swiftly in compact flocks, screaming loudly. They can be very destructive in fruit orchards and ripening corn.

Broadbills are birds of dense, tropical forest. They feed mainly on fruit and make large, untidy nests, nearly always in bushes overhanging water, where they look very like the flood debris that gets caught up in riverside vegetation. The green broadbill comes from Malaysia and Indonesia.

Top: The red junglefowl is widely distributed in Asia and the male, shown here, is a very handsome bird; the hen is much more soberly coloured. Junglefowl of this and possibly some other species are the ancestors of the domestic fowl.

general confusion of a lot of birds flying in an unpredictable zig-zag fashion the raptor is much less likely to be successful in its aim; and a large group may turn in force to mob and drive it away.

The decorative sunbirds occupy a specialist niche, feeding on nectar. They flit from blossom to blossom of the various flowering trees and shrubs, extruding their long, tubular tongues to extract not only the nectar, but also the small insects the blooms attract. In this way they help to cross-pollinate the various plants. Red flowers with big trumpets are typically pollinated in this fashion. The birds usually perch on a convenient twig or stem to sip the nectar but they can hover on rapidly vibrating wings in front of a flower, though not for long.

The forest floor provides a habitat for several important groups of birds, important not only because they play a major part in the forest ecosystem but also because they have provided the ancestors of several of Man's domestic species. Chickens, the most abundant birds in the world today, roam wild as jungle fowl in the Asian forests as do colourful, exotic pheasants and peafowl.

One of the most familiar sounds of the rainforest of South-East Asia is made by that most rarely seen bird, the great argus pheasant. The male clears the litter from the forest floor to make a bare ring about 7 metres across. For most of each morning he sits in or paces about his ring, clearing away any leaves or seedlings. Every few minutes he gives the clear double *kaūwāu* call which is answered by cocks on neighbouring rings and echoed round the jungle. Any female attracted to the calls is treated to one of the most spectacular displays of the bird world. The male struts to and fro, pecking rhythmically at the ground. Suddenly he flashes open his wings and the enormous secondary feathers fan out, completely obscuring his head. The long tail is raised erect. The cock rattles his raised wing feathers, decorated with rows of 'eyes', and then lowers them slowly.

This display has rarely been seen in the wild, even by the most enthusiastic naturalists. The argus pheasant is extremely elusive. His incredible eyesight and sharp hearing enable the cock to slip away down one of his escape routes at the least sign of potential danger. Away from his dancing ring the male gives a different call, a single clear *wow* repeated with increasing tempo a dozen times or more. According to the natives, the female pheasant utters a similar cry.

Pheasants, and the closely-related partridges, scratch among the leaf litter, feeding on fallen

63

fruit, insects and termites. Since they are all ground-nesting birds the females tend to be rather drab in appearance to blend with the background. The cocks, however, are extremely handsome, with brightly coloured plumage. The male fire-back pheasant is a resplendent creature with blue, green and red feathers and a spectacular white tail. Occasionally the males fight out disputes, leaping high and stabbing at their opponents with their spurs. Normally, however, these birds are seen wandering in small parties, usually with only one cock in the group.

Among the other ground-feeding families the pittas, too, are insect eaters. This gaily coloured group is well represented in the Oriental region. Pittas are plump, short-tailed, long-legged birds, able to move about with kangaroo-like hops. Their bright plumage makes them unusual among forest species. The jungle babblers, for instance, are nondescript birds, flitting through the undergrowth in search of insects, either singly or in small groups. In those species which do sport gay colours they are usually confined to the male while the female is much more cryptic, even when the nest is a closed ball-like structure as in the pittas. The obvious disadvantage of bright plumage is that it

draws the attention of would-be predators. For such a character to have evolved it must confer some advantage that outweighs this drawback and the usefulness of such advertisement probably lies in social recognition, territorial disputes and the attraction of mates.

Although the environment within the rainforest remains fairly constant, there is a marked seasonality of rainfall. Moreover almost every species of tree is seasonal in its flowering and fruiting. Although it is true to say that one can find flowers and fruit in any month of the year, there are definite times when the majority of species produce their crop. This is reflected in the fluctuating availability of insect types. It is not surprising, therefore, that the majority of forest birds show seasonality in breeding and nesting, correlated with the availability of their food supply.

At certain times of the year, corresponding with the winter months of the temperate regions, the avifauna of the forest is supplemented with an influx of migrants, some arriving to overwinter, some on passage to Australia. Most of the winter visitors, like the hawk cuckoo, blue and white flycatcher, crow-billed drongo, arrive from the Palaearctic to the north but a few species, including

ROSS

Left: The great argus pheasant is at the same time one of the most sombre and most magnificent of all pheasants. In the nuptial display the male raises his wings so that the secondary flight feathers form a vertical screen ornamented with rows of spots, each of which is colour-shaded to appear spherical. Argus pheasants are found only in the primary rain-forest of South-East Asia and, though not rare, are seldom seen.

Wagler's pit viper is one of a group of venomous snakes which have a pit between each eye and nostril containing an organ sensitive to very small amounts of radiant heat. Using this they are able to find warm-blooded prey, small mammals and birds, in complete darkness.

The largest scorpions are those found in tropical rain-forest, where they hide by day under fallen trees and branches and come out to hunt their insect prey by night. Greenish black in colour, they may be 15 centimetres long from head to tail.

the bronze cuckoo and the long-legged pratincole, reach South-East Asia from their breeding grounds in Australia. Not all the birds that winter in the tropics are forest species – the pratincole, for instance, is more at home in a paddy field – but many must pass through the forest belt *en route* to a more suitable habitat.

The forest at night

Night falls swiftly in the tropics and dusk comes even faster in the shady forest. The end of the day is heralded dramatically by the weird wailing of the katydid crickets. Nightjars circle in the darkening sky, hawking midges and calling plaintively. Leaf monkeys settle in their sleeping posts, diurnal squirrels hurry to reach their dreys and the

MACKINNON

creatures of the night emerge from their holes.

The flying squirrel stretches his cramped limbs and sprawls in the rays of the setting sun, grooming his fur before he glides off in search of his evening meal. Owls blink, their sharp eyes scanning the surrounding forest for any unwary prey. The black scorpion scurries over the leaf litter, his poisonous tail-sting raised high. Large, hairy spiders sidle off in search of small victims and the geckos emerge from cracks in the tree bark and take up their feeding positions, ready to snap on moths and flies that land nearby. The little *chikchak* gecko, frequently heard in the jungle, has also become a common inhabitant of human dwellings where it scurries up and down the walls, chattering furiously.

Many of the forest snakes hunt at night. The yellow and black-banded kraits and cat snakes, colourful coral snakes and enormous king cobras feed exclusively on other snakes. The common cobra and pit viper hunt rats. The pit viper is equipped with heat receptors on the front of the head which enable it to detect the presence of warm-blooded prey. All of these species are poisonous though this is not typical of snakes as a group. Snake venom contains several active ingredients and serves not only to kill the victim, either by impeding respiration or destroying the red blood corpuscles, but also acts as a digestive juice.

Rats and mice abound in the rain-forest and are found even in the top layer of the canopy. South-East Asia is the original home of this rodent group. They are cousins to the nocturnal porcupines; in fact the long-tailed porcupine, with its short, flat quills covered in hair, looks just like an enormous rat. The Malay porcupine, however, has an effective armoury of sharp spines. When threatened, it raises its quills and, rattling them together angrily, charges backwards into the face of its enemy. Porcupines feed mainly on roots and woody material but their gnawing teeth are sufficiently strong to tackle the bones of dead animals. Even elephant tusks lying on the forest floor may bear the scars of porcupine incisors.

The porcupine is not the only animal in which hairs have been modified to form protective armour. In place of spines, pangolins, or scaly ant-eaters, have evolved an almost impenetrable coat of overlapping horny plates. When disturbed, the pangolin wraps its tail over its head and rolls up into a tight ball so that any potential predator is confronted by an armoured sphere which it finds

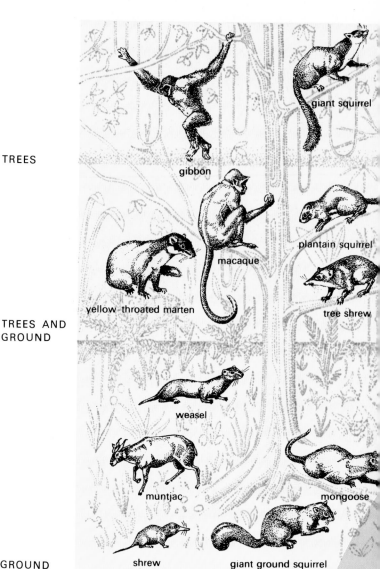

DAY

TREES

gibbon

giant squirrel

macaque

plantain squirrel

yellow-throated marten

TREES AND GROUND

tree shrew

weasel

muntjac

mongoose

GROUND

shrew

giant ground squirrel

extremely difficult to prise open. The Malay pangolin is widespread in South-East Asia and another species is found in Ceylon. During the day these animals hide in holes under trees but at night they wander off in search of termite and ant nests. With powerful legs and sharp claws, they rip open these colonial nests and probe the galleries with their long thin tongues. Eggs and pupae adhere to the sticky surface and are drawn back into the toothless mouth. Since they are agile tree climbers and can hang by their prehensile tails, pangolins are able to tackle tree nests just as easily as those in the ground or under logs. The tail has yet another use, for young pangolins ride like eager jockeys on their mother's tail until they are big enough to fend for themselves.

Small, insect-eating shrews scurry among the

leaf litter. Their musky odour makes them un-attractive to many raptors. The largest of all the insectivores, the moon rat, is so evil-smelling that it need fear no predator and can afford to advertise its presence by its noisy shuffle and bold markings. In Borneo these creatures are yellowish-white and stand out clearly against the dark background of the forest floor. In Sumatra and on the mainland moon rats are equally conspicuous, with grey bodies and black and white striped faces. The long, sensitive nose, terminating in a curious pink rosette, is used to sniff out earthworms and arthro-pods, and occasional molluscs. The moon rat at night occupies almost exactly the same niche as the common tree shrew uses by day.

In the same way the pen-tailed tree shrew occupies a similar niche to that of the lesser tree

Different animals are active in different parts of the forest at different times of the day. This diagram shows how Bornean mammals can be arranged into horizontal strata. The canopy is utilized by a variety of small, agile creatures, split clearly into those with diurnal or nocturnal habits. The forest floor is frequented by the largest mammals as well as some of the smallest. On the forest floor the division into day and night animals is less clear. Many of the large herbivores like the elephant, deer and rhinoceros are active both by day and night.

shrew by day. It darts among the lower branches of the forest catching insects and can usually be found among dense clumps of vines which provide lateral travel routes. Very occasionally it descends to the forest floor. The pen-tailed tree shrew has the big round eyes we associate with nocturnal mammals and gets its name from its strange, naked tail which sprouts a flat tuft of white hair near the tip.

Even more unusual in both appearance and behaviour is the tarsier. It has such enormous eyes that it cannot revolve them in their sockets and must turn its whole head to look from side to side. Its big eyes and ears give it remarkable vision and hearing and enable it to detect insects scuttling over the forest floor. Having once marked its prey, it leaps with great precision and devours the catch greedily. Tarsiers are equipped for their role as

Weighing up to 8 kilograms, and with an effective armoury of sharp spines, the Malayan porcupine is well able to take care of itself in the forest. It digs extensive burrows and emerges at night to feed on starchy roots – including the cultivated varieties such as tapioca. It is also fond of bones, which it rasps away with its strong front teeth.

vertical, clinging leapers with long, spindly arms and legs and broad, flat pads on the fingertips. These pads resemble those of the tree frogs and serve the same purpose. They give immediate, good contact with twig and leaf surfaces as the tarsier leaps from one vertical support to the next.

Using quite different tactics, lorises climb slowly and cautiously through the branches, catching insects by stealth rather than speed. The slender loris is found only in Ceylon but the chubbier slow loris occurs throughout Indo-China, Malaya and the Sunda Islands. Both species feed on a mixed diet of fruit and young leaves, crickets and moths and will even raid birds' nests for eggs and nestlings.

Like the tarsier, the lorises are primates, related to lemurs, monkeys and apes. As is typical of this order they have flat finger nails rather than claws. The lorises, however, have retained, or possibly re-evolved, one long claw on the second toe of the hind foot. This is a grooming claw for raking through their thick fur and scratching behind the ears during the nightly cleaning session.

Although the thick canopy blots out the faint light of the moon and stars, the jungle night is not completely black. The forest floor glows with an eerie, living light. Fungal hyphae and bacteria on the decaying leaves and clumps of luminous toadstools shine phosphorescent green. Bioluminescence is not confined to the plant kingdom. Whereas many of the larger night animals maintain contact and establish range boundaries by clear, carrying calls or scent marking, several of the insect groups rely on light signals for communication.

Glow-worms weave among the leaf litter. Adult fireflies flash through the canopy, winking out their sex and species to prospective mates. Larval fireflies also glow, but for less obvious reasons. Most have just two glowing points near the tail but one curious specimen has two complete rows of green beacons so that it looks like a tiny train chugging through the jungle night. All the larvae of this

Tarsiers are primates, relatives of monkeys, apes and Man. Only 15 centimetres long, they live among trees and bushes in much the same way that a tree frog does, clinging with their adhesive toes and making long standing jumps from branch to branch. Tarsiers are nocturnal and predatory, catching insects and other small animals. They live only on some of the large islands of South-East Asia; this is the western tarsier, found in Borneo and Sumatra.

Lorises are small primates that climb slowly and deliberately about in trees and bushes at night, resting and sleeping during the day. They feed on insects, lizards and small birds, catching them stealthily in the dark; they also eat fruit. Like other primates, they have flat finger nails rather than claws, but they also have a special long claw on the hind foot for self grooming.

The slow loris (above) is quite widely distributed in eastern Asia, but the smaller, spindle-legged slender loris (right) is confined to south India and China.

BEAMISH

BOARDMAN

70

beetle family feed on slugs and snails, piercing the rubbery bodies with their long, slender mouth-parts. The large trilobite larva resembles an ancient trilobite fossil. Nobody has ever succeeded in catching or breeding an adult of this species and it may well be neotenous, and remains in the larval form throughout its life.

The ecology of the forest at night, though less well understood, is just as complex as by day. The lives of herbivores, frugivores, insect-eaters and carnivores are all intricately interwoven in every strata of the forest. Some of the large terrestrial animals such as elephants, tigers, pigs and deer are active both by day and night but the arboreal fauna is more clearly divided into diurnal and nocturnal animals. Some night-time creatures, like the pangolin, exploit niches left vacant during the day but many nocturnal animals such as the moon rat and the pen-tailed tree shrew utilize the same resources that other rain-forest species have used during the day-time.

Pangolins look more like reptiles than mammals but their over-lapping scales are really formed of hair, and the underside of the body is hairy; when they are threatened they curl up like hedge-hogs to protect themselves. They are quite toothless and feed on ants and termites just as the anteaters of South America do, though pangolins and anteaters are not closely related. Being agile climbers they can raid tree nests as well as those on the ground, ripping them open with their sharp claws, and probing with their long sticky tongues.

The Indian darter is a diving bird rather like a cormorant. It hunts fish underwater, and can often be seen perched on a branch with its wings spread, drying its soaked plumage in the sun.

The Fringe Habitats

WAGER

The coastal flood plain of a river in Borneo. Streams wind and meander in wide loops, joining each other or forming ox-bow lakes. The flat land between is covered in thick swamp forest and is liable to frequent flooding.

The false gavial is a crocodile found only in Borneo, Sumatra and the Malay Peninsular, where it lives in large inland swamps. Its long, slender snout and small teeth resemble those of the true Indian gavial, but this is probably an adaptation for catching fish and not an indication of close relationship.

BANKS/NSP

Around and within the tropical rain-forests of Asia are habitats sufficiently different in their ecology and animal life to merit separate consideration. These fringe habitats include the great limestone outcrops that pierce the green monotony of the forest, the twisting brown rivers that drain these moist regions and, of course, the coastal swamps and beaches that divide the jungles from the sea.

The riverine animals

The broad rivers that cut brown, meandering highways through the monotony of dense, green jungle, provide a habitat for a very different fauna from that of the tropical rain-forest. Wash-off after torrential rain leaches minerals and nutrients from the soil of the forest floor and these are carried by the muddy tide down to the river's mouth, providing a constant and plentiful food supply for river-living animals. The banks, too, are a rich environment. Open to the sky and receiving precious sunlight, they support a wealth of ground herbage that is found nowhere else in the forest.

The rivers through the rain-forest contain numerous fish that feed on the humus and debris that swirl down in such abundance. Smaller

Large river monitors may grow to nearly 3 metres in length. They have flickering, forked tongues rather like snakes and, also like snakes, they swallow their food – fish, insects, fledglings and carrion – whole.

tributaries, where the waters are clear and light and nutrients are readily available, support blooms of green algae, another food source for shoals of tiny fish. Three-spined catfish lurk beneath fig trees, waiting to gobble up falling fruits. Large, predatory *tarpa* cruise through the shallows, preying on the smaller fry. Even sting rays and sawfish wander far up these rivers, miles from the sea, to feed on the shrimps and crustaceans that live among the gravel.

Big, soft-shelled turtles lie submerged on the muddy bottom, their razor sharp beaks ready to snap on any unwary passer-by. During the summer months these turtles, or *labi-labi* as the natives call them, clamber out onto sandy spurs to lay their leathery white eggs. The eggs of the labi-labi are regarded as a great delicacy, not only by the river-dwelling humans of South-East Asia but also by the scavenging monitor lizards. River monitors may grow up to 3 metres long. Their forked snake-like tongues can smell out death and decay in the water or neighbouring jungle and they home in on rotting corpses to feed. They also catch fish and insects and prey on birds' nests and fledglings.

Larger and even more ferocious are the crocodiles. At the turn of the century these were extremely common on most of the rivers of tropical Asia. They took such a heavy toll in human lives that successive governments offered bounties for the feet and eggs of destroyed animals. More recently, however, lucrative rewards have been offered as a result of the fashion for reptile-skin shoes and handbags. High prices are paid for skins and even greater sums for clutches of eggs. Crocodile farms have sprung up in Singapore, Sumatra, Thailand and elsewhere but, since it is extremely difficult to breed crocodiles in captivity, they rely on clutches of wild eggs to build up their stock. The result has been a terrible decline in wild crocodiles, which have become almost extinct in many countries.

The large Asiatic crocodiles feed on fish as well as on the occasional monkey or deer that strays too close to the river's edge and is knocked, dazed and floundering, into the water by a swish of the reptile's powerful tail. In Borneo there is a super-abundance of food for the crocodiles during those months when migrating bands of bearded pigs ford the wide rivers. At these times the marauding killers grow fat on the juicy rumps and bellies of the stragglers and there are plenty of left-overs for the opportunist monitor lizards. The false gavial, a rather rare, specialized type of crocodile, feeds

GRANDJEAN

almost entirely on fish. As it swims through a shoal of fish it lashes its long, slender snout from side to side, snapping at its prey, which it swallows whole. It poses no threat to man but has been another unfortunate victim of the flourishing handbag industry.

Otters, too, make a living from the schools of fish. These sleek animals are expert swimmers and they are a common sight in rivers and estuaries, travelling alone or in playful family groups. Because of their expertise at snatching fish from nets they are hated by the local fishermen. The rarer otter civet is another semi-aquatic mammal with similar eating habits. This curious creature with its hairy snout and otter-like webbed feet is a member of the primitive family of civet carnivores which had established a specialized niche before the true carnivores of the cat and stoat families reached South-East Asia.

Several large herbivores come down to the river's edge to browse and graze on the luxuriant vegetation. Wild pigs root among the grass and ferns and the common sambar deer feed here at night. Sambar are the largest deer of South-East Asia, often weighing over 150 kilograms. The stags bear neat six-pointed antlers. Another nocturnal visitor to the river banks is the banteng. These wild cattle hide up in secondary forest or the forest fringe during the day and venture out to graze at night. They are splendid beasts. Bulls are solidly-built, powerful animals with thick, curved horns and coats of glossy black with white socks and a white rump patch. The smaller cows and calves are a uniform chestnut colour. In Malaya the closely related, rangier seladang has similar habits. Large herds of a dozen or more make regular pilgrimages to water holes and sites where they can obtain mineral-rich soils.

The resident monkeys of the river banks are the long-tailed macaques. They can live in the rainforest, and even in mangrove swamps, but they are essentially riverine monkeys and can often be seen scampering along the shoreline catching crustaceans, despite the danger from eager crocodiles. This pastime has earned them the alternative name

Freshwater fish are an important source of food for the people who live along the courses of the rivers running through forest. Catfish, ugly creatures with huge mouths and long whiskers or barbels, make excellent eating.

Sambar are the common large deer of southern and eastern Asia, and are found in forest of all types. Like other herbivores, they come to the river banks to feed on the rich grass. The antlers of the stag may measure 60 or 70 centimetres along the outer curve, and are shed periodically, not necessarily every season. The hind, like almost all female deer, is without antlers.

of crab-eating macaques. Like other higher primates they are very fond of fruit and crowd into the overhanging fig trees to enjoy the sticky crop. Long-tailed macaques are good swimmers and can travel several metres underwater. They live in large groups of twenty or more individuals and have complex social hierarchies with great jealousies and squabbling over status among the males. During the day the troops move away from the water to forage in the forest for leaves, fruit and particularly flowers, but they always return to their riverside sleeping trees by the water at night. Their raucous alarm call has lead to their Malay name of *kera*.

The great reticulated python is never far from water. During the day these enormous serpents curl up in dark holes in river-banks, hollow trees, under fallen logs or in rock crevices. At night they emerge to hunt. Big specimens, over 7 metres in length, can easily overcome wild pigs. Striking with open mouth to deliver a hammer-like blow, the python first knocks its prey senseless, then winds its coils round the unconscious animal and proceeds to crush it to death before devouring it whole, head first. Like most other snakes, the python can dislocate its jaws to allow it to swallow prey much broader than itself. Grossly bloated, it returns to its hole to digest the meal and it may be weeks before it feels the need to hunt again. Pythons are rarely seen by the forest traveller but they are undoubtedly quite common, as is revealed during the heavy rains. When the great rivers flood many small pythons are washed out of their holes and climb up into overhanging branches where they can be found in fair numbers, coiled above the water.

RAO/NHPA

The seladang or gaur is the largest of all wild cattle and is distinguished by the white 'stockings' and the hump-like ridge over the shoulders which is about 2 metres above the ground in big bulls. Seladang are found in both deciduous and rain-forest in eastern Asia, living in herds of a dozen or more animals. Adults of both sexes are black, but calves up to six months old are golden bay coloured.

Long-tailed or crab-eating macaques are the familiar monkeys of southern Asia eastward from Burma and Thailand. They inhabit all kinds of country, from hill forest to coastal mangrove. Though in mangroves and riverine areas they do catch crabs and other crustaceans, the name 'crab-eating' gives a false idea of the species as a whole; they will in fact eat practically anything. Near towns and villages they often become bold and tame.

MORRIS/ARDEA

79

The Brahminy kite now lives almost wholly as an associate of Man on the coasts of tropical Asia. It frequents sea ports and fishing villages, feeding by scavenging offal and waste thrown overboard from ships. It will also catch frogs, crabs and young domestic chickens.

Top: Darters are less buoyant than other water birds, and can float with their body submerged and only the head and neck above the surface. Underwater they hunt with neck flexed and ready to be shot forward, the bill carried like a pointing spear, ready to transfix or seize the fish.

The riverine habitat supports its own bird life with, as one would expect, many species feeding on fish. This is the hunting ground of the great raptors, the fish eagle, osprey, fishing owl and the Brahminy kite, which takes large crickets as well as fish. The Brahminy kite is an attractive bird, common throughout tropical Asia. The Sea Dyaks of Borneo regard it as the earthly manifestation of one of their foremost deities and other tribes read significant omens into its behaviour.

Herons stand in the shallows spearing their prey. Snake-birds or darters hunt under water. They are voracious eaters of fish and are capable of diving under water for minutes at a time. They swim quickly, following the shoal's every turn until they get close enough to strike a fish with their sharp lance-like bill. The darter's subaquatic skill is due in part to the absence of the greasy oil with which most birds waterproof themselves. As a result, its feathers become waterlogged, making it less buoyant than other waterbirds. It is able to float with its body submerged and only its neck and head breaking the surface. Darters are frequently seen perched on branches overhanging the water, their wings outstretched in the sun to dry out the soaked plumage.

Gaudy kingfishers flash to and fro. They dive from favoured perches to catch small fish and carry them back to secure holes in the riverbank where their hungry nestlings wait. Blue-throated bee-eaters also nest in river banks but in large, noisy colonies. Bee-eaters leave their perches to make lightning dashes to snap up any bees, crickets or beetles that catch their eye. They return with the prize held tightly in the beak and smash it against a branch to break through the thick cuticle before the insect is swallowed.

Tailorbirds hunt energetically among the bushes, rushing along branches with surprising speed to snap up tiny insects. The red-tailed tailorbird is a neatly coloured, active bird with an upcocked red tail and perky red cap on its head. Each evening the cocks sing, giving their cheerful *chick chiki* call, vigorously chasing other birds off their territories.

The reticulated python rivals the South American anaconda for the distinction of being the world's biggest snake: adults grow to almost 8 metres, though they are mature and able to breed at 3 metres or less. Large pythons prey mainly on mammals up to the size of monkeys; young ones are useful destroyers of rats.

Kite swallowtails (*Graphium* of several species) and a sawtooth (*Prioneris*) cluster by a river in Sumatra. In tropical Asia groups of butterflies may often be seen on riverside sand or shingle. It has been shown that they are usually attracted there by animal or human urine, apparently craving the organic salts which it contains. Curiously, in most groups of butterflies only males are found at these gatherings.

The magnificent Rajah Brooke's birdswing is found in Malaya, Sumatra and Borneo. It was discovered in Sarawak in 1855 by the English explorer A. R. Wallace, and named in honour of the 'White Rajah' who then ruled the territory. Male butterflies gather, sometimes in large numbers, at hot springs whose water contains mineral salts. The females live in the canopy and are seldom seen.

Tailorbirds nest along the river banks or in clearings on the forest fringes – even in town gardens. Nesting is a secretive affair. Adults travel by such devious routes that it is extremely rare to find a nest even though the birds themselves are not at all uncommon. The nest is made by sewing two or more leaves together – occasionally the birds may sew the opposite edges of a single, large leaf. To achieve this remarkable structure the tailorbird uses its beak to pierce neat holes through the two layers and draws them together with threads collected from spiders' webs. Within this 'sewn' tube it builds a grassy cup-shaped nest in which to lay its eggs.

Among the most spectacular sights of the riverside are the carpets of fluttering butterflies attracted to sulphur and mineral springs or the damp sand where some animal has recently urinated. They circle and alight to sip delicately at the salty moisture with their long proboscides. Blue *Graphium* butterflies, purple-brown milkweeds, and smart black and white swallowtails mix in a swirl of colour with the cream and yellow of the sulphur and *Eurema* butterflies. Drifts explode into snowstorms as one draws near and white butterflies chase one another in dancing chains along the river's edge. Most magnificent of all are the Rajah Brooke's birdswings with their wings of velvety black and iridescent green. Only the males have time to gather to drink at the water's edge; the equally resplendent females spend their time fluttering high in the canopy, looking for suitable places to lay their eggs.

Limestone caves

Calcium carbonate is secreted by several marine organisms, bacteria, brachiopods, crinoid lilies and, of course, corals. Normally it remains dissolved in the sea water but it is chemically rather unusual as it becomes less soluble with increasing temperature. In the warm tropical seas where many of these organisms live the water becomes saturated and calcium carbonate precipitates out as a white solid. As it sinks into the cooler waters beneath it dissolves once again but where the sea is very shallow it settles as a white ooze which gradually crystallizes into sedimentary limestone.

The Sundra Shelf was a shallow tropical sea for many millions of years before tectonic movements thrust up the land masses and islands that shape South-East Asia today. As a result great ranges of limestone hills and huge isolated boulders and limestone outcrops are found scattered throughout the region. These are still being shaped by the elements. Falling rain absorbs small amounts of carbon dioxide from the atmosphere to form dilute carbonic acid which slowly dissolves the limestone. Surface streams gradually wear away the rock along the weak fissure planes, carving out gullies and eventually disappearing underground to flow through hidden galleries eroded in the bedrock. Where the streams reappear are chains of caves. These labyrinths are fascinating places with weirdly carved buttresses and dripping stalactites and stalagmites but they are of particular interest because they support a characteristic fauna quite different from that of the surrounding forest.

Caves are dark places devoid of sunlight and,

In Malaya and Borneo steep limestone hills rise here and there from the rain-forest and many contain caves produced by solution in percolating water. The caves are sometimes large and spectacular, with stalagmites and stalactites. They are the home of innumerable bats and of the small swiftlets whose nests of coagulated saliva are relished by Chinese gourmets.

apart from the moss and ferns that grow around the mouth, they support no green plants. They owe their abundant animal life not to their attractiveness as feeding places but because they offer comparatively safe shelter for enormous numbers of swiftlets and bats. Bats by night and swiftlets by day spread out far and wide over the surrounding forest to feed but return to roost and breed in the caves. The other animals of the caves depend, directly or indirectly, on the bats and swiftlets, living off them either as predators or parasites or as part of the food chain supported by the piles of powdery brown guano, rotting corpses or insect debris that they drop.

Many species of bats live in limestone caves; most are insectivorous or predatory, feeding on smaller bats. They have a highly developed system of sonar echolocation which they use to find their way through the dark cave galleries, as well as for travelling and locating food in the forest at night. The leaf-nose bats and horseshoe bats are typical examples. In the former leaf-like noses give a grotesque appearance but are vital for controlling and beaming their ultrasonic emissions. Both have large, curiously shaped ears, the shape of nose-leaves and ears being quite characteristic for each species. Such elaborations produce highly individual and efficient codes of sonar emission and reception so that the bats can recognize members of their own species and their own personal echoed signals from the barrage of other bat squeaks that fill the forest night.

The great flying foxes are not cave dwellers but some of the smaller fruit bats do roost in caves by day. The commonest of these is the cave fruit bat which, despite its name, feeds mainly on nectar. As these bats do not possess echolocation they are confined to the more open caves where they have sufficient light for their manoeuvres. Of the fruit bats only the rousette bats can fly in total darkness for they have developed echolocation. Their high-pitched buzz of tongue clicks is less efficient than the sophisticated ultrasonic, laryngeal emissions of the insectivorous bats, but is quite adequate for their needs.

Caves are also the roosting places of hairless bats. These foul-smelling, naked, wrinkle-skinned bats are among the world's ugliest creatures. As if this were not sufficient they carry, exposed for all to see, the largest and most unattractive of parasites, an enormous hairy earwig that feeds on the scurf of their skin. This earwig is probably the biggest ectoparasite ever found on a mammal; it is about a quarter of the body length of its host, which is equivalent to a man permanently encumbered with a lobster! Another rare species of hairy earwig is found in association with the hairless bats. It lives beneath the roosts, scavenging on the bats' pungent droppings, a food source which few other creatures find attractive.

Young bats are born blind and naked. For a day or two they may be carried by the mother but after this they are left hanging in the roost while she collects food in the forest. Those unfortunate enough to fall from the roost flutter helplessly for a few seconds before being submerged under the tide of scavengers that scour the guano.

Cave swiftlets also breed in the caves, building cup-shaped nests up the sides and in the dome of the rock cathedrals. Nesting is seasonal, varying according to species and from one cave to another. Large colonies of one or two million birds may roost and nest in the same cave but they are excellent fast fliers and hunt over an area of many square kilometres of forest. At dawn, as the bats stream home, the swiftlets pour out of the caves in a great surge but for most of the day the outflow equals the numbers of those returning. In the evening there is a second great outpouring and the last swiftlets do not return until well after dark. Their ability to fly in the dark and find their nests and roosting sites is, like the bats, dependent on echolocation. The swiftlets begin to emit high-pitched, rattling calls as soon as they enter the caves and navigate their path according to information received from the bouncing signals.

Swiftlets have very short legs, quite unsuitable for settling on the ground or taking off again. They remain permanently airborne except when clinging to their rocky ledges and nests. They are consequently unable to collect mud to build their nests like other swifts and swallows. Instead the birds secrete saliva and felt it together. It is this hardened saliva which is used in Chinese cookery to make the much-prized birds' nest soup. An important industry has developed around the collection and preparation of the swiftlets' nests.

It is probably safety from predators that recommends the cave walls and roof to the bats and swiftlets. The only predator that can reach them there is a large wingless cricket which feeds on the young and can pierce the birds' eggs to suck out the contents. Several vertebrate predators, the marbled cave snake, civets, rats and monitor lizards, have made their home within the caves and feed on young birds and bats that fall to the ground. Predatory

birds, such as Brahminy kites, peregrine falcons and the crepuscular bat hawks wait outside the caves to attack bats and occasional swiftlets as the flocks fly in and out. The bats and swiftlets have their own attendant parasitic community, including blood-sucking bugs, ticks and flies. The larvae of the moth *Pyralis pictalis* devour the saliva nests of the swiftlets, weakening the structure and sometimes causing them to crash to the ground.

By far the greatest loss to the swiftlets, however, is caused by Man's nest-collecting forays. Chinese traders first came to Borneo to buy birds' nests over a thousand years ago. Gradually the trade died down as fewer and fewer Chinese boats came south but it revived again during the nineteenth century.

The nests most commonly collected are those of Low's swiftlet. These nests are termed 'black' because of the feathers incorporated in the saliva. Careful cleaning is necessary before the nests can be cooked and they are consequently less expensive than the 'white', almost pure saliva, nests of the brown-rumped swiftlet which are, however, far less common. The nests of the grey-rumped swiftlet are also 'white' but these birds only roost in small colonies in coastal caves and their nests cannot be collected in large numbers.

The birds already mentioned roost in the inner-most dark recesses of the limestone caves but the white-bellied swiftlet, which does not possess echo-location and must find its way by sight, nests on ledges in the light entrances to the caves, in house eaves and other suitable places. Its nests incorporate so much moss and other vegetable material that they are commercially quite useless.

Nests are collected only at certain times during the breeding season: once very early, since birds will often rebuild, and again towards the end of the season when many of the fledglings will have left the nest. Even so there is usually a considerable loss of young and eggs and there can be no doubt that the collection of such enormous quantities of nests does reduce the number of fledglings that can be reared. Despite this predation the colonies in the various caves show no decline in numbers.

Collecting methods vary from cave to cave. In some places nests can be reached with a pole from the ground. Elsewhere the collectors erect elaborate ladder networks to enable them to reach the highest ledges. Perhaps the most terrifying method of all is that employed in the famous Great Cave of Niah where Punans climb vertical poles of jointed bamboo that stretch 50 metres up into the dark vaults. They use a long pole to which they attach a

LIM BOO LIAT

knife for cutting off the nests and a candle to light the operation. Every now and again someone falls and dies but so long as the price of nests remains high there will be Punans willing to make their living this way. Perhaps the most ironic feature of the whole business is that when it has been thoroughly cleansed and cooked the swiftlet's nest is quite tasteless. It is used solely to give texture to some other fruity or meaty dish.

Bat and swift guano is also collected by man for fertilizer. Since it is collected at a much faster rate than it is deposited, this trade has rather a limited future. The natural exploiters of the guano are the cockroaches *Pycnoscelus*. These are found in enormous numbers, up to 3,000 per square metre, devouring the droppings and any eggs, fledglings or bats that fall among them. Also feeding on guano are the little, sand-cased caterpillars of the moth *Tinea palaeochrysis*, beetles, flies and innumerable springtails.

The insects active on the guano are only one link of a food chain; they, in turn, are caught and eaten by a multitude of predatory species. Spiders lurk in the crevices, spinning their webs to trap the golden cockroaches. One representative of the *Liphistius* genus of spiders, which can be regarded as living fossils because their bodies are segmented like those of their arthropod ancestors, is found among the rocks outside the Batu Caves of Malaya. Secure in its silk-lined tunnel within a rock crevice, it sits behind its trap-door waiting for an insect to touch one of the silk threads that converge upon its home. A change in tension of the thread warns the spider of the approaching prey and it hurries out to seize the trespasser and overpower it.

LINDGREN/ARDEA

Rousette bats are the only fruit bats that orientate in darkness by echo-sounding as most of the insectivorous and predatory bats do. The sonar of rousettes is much less refined and efficient than that of the insect-eaters, and the sound-pulses are produced in a different way. This is Geoffroy's rousette.

The white-bellied swiftlet is a small relative of the swift. It frequents the limestone hills of Malaya and Borneo, attaching its nest to the rock in well lighted areas at the mouths of caves. It is closely related to the swiftlets whose nests are collected from further inside the caves and used for birds' nest soup.

Whip scorpions are uncommon in the forest but they abound on the floors of the caves, feasting on springtails. True scorpions catch unlimited numbers of cockroaches and the fast-running, long-legged centipede *Scutigera* will tackle any insect that comes within range. These centipedes have long, thread-like antennae and elongated sensory rear legs so that they can move through the inky blackness sensing their prey. Small groups of darkling beetles scuttle among the carpets of writhing cockroaches, collecting and consuming any corpses they come across.

Another group of predators that frequent the caves are the ant-lions. In drier parts of the tropics their conical traps are a familiar sight but in the tropical rain-forest they are rare. Ant-lions can only breed in areas sheltered from rain since the larvae require dry, dusty sand or soil in which to make their pits. Dry, light cave mouths provide ideal conditions as does the ground beneath the traditional stilted Malay and Dyak houses. The larva, a segmented sac bearing a small head armed with enormous curved jaws, lives under the dust. It travels backwards and forms its pit by spiralling round in ever-decreasing circles, flicking out the loosened sand with its head. The result is a steep-

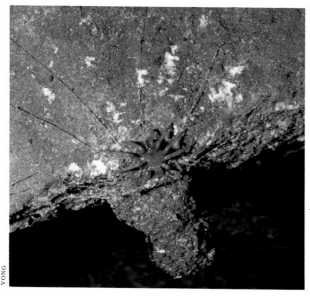

YONG

In Malaya and some neighbouring territories there are spiders belonging to an exceedingly ancient and primitive group; very similar fossil spiders have been found dating from the Carboniferous period, some 300 million years ago. The entrance to their burrows is closed by a 'trap door' and on the surface, threads of silk radiate outwards around it. When prey approaches, the thread tension changes, and the spider emerges to seize it. This species, *Liphistius batuensis*, is found at the Batu Caves in Malaya.

sided, conical hole with the larva hidden out of sight at the bottom, its jaws open ready to seize any insect that might fall into the cavity. Trapped insects struggle to climb out up the loose, sliding dust but invariably slither back down into the pit. The ant-lion's snapping jaws carve into the prey, dragging it under the sand where it is pierced and its body juices are sucked out. The discarded empty shell is flicked away and the ant-lion makes a new trap. Adult ant-lions are delicately winged creatures with slender abdomens, very similar in appearance to damsel flies. They fly by night but can be observed by torchlight hovering gently above the dry soil, their down-curved abdomens depositing eggs beneath the surface.

The ecology of the limestone caves has been a subject of enormous human relevance ever since Man arrived in South-East Asia. Long before the development of agriculture, Stone Age Man, like the ant-lion, found the dry sheltered cave mouths were the best place to live. Remnants of these early communities have been found in the lower cave levels: stone adzes, flaked and ground knives and axes, barkcloth beaters made from carved horn, as well as the animal bones and mollusc shells that tell us about Early Man's eating habits. Today caves are still regarded with awe as spiritual or haunted places. Many tribes throughout the region use them for burial grounds or as shrines and places of worship. The Batu Caves in Malaya house an Indian temple which is visited by processions of Hindu pilgrims during the festival of Thaipusam.

The sea coast

For thousands of miles the coasts of the Oriental region twist and turn round bays and peninsulas, great islands and tiny coral reefs. In places heavy seas smash hard against jagged rock; elsewhere waves lap gently on white, sandy beaches or swirl and gurgle among the tangled mangrove roots that grow on the broad muddy margins between land and sea.

Where the seas are shallow, corals build delicate calcium carbonate colonies and colourful fish hide among the branching forms and anemone tentacles. Poisonous sea snakes weave sinuously through the water, propelled by the paddling of their flattened tails. Occasionally the sea bed is darkened by the shadow of a passing shark or turtle. Turtles are an ancient group of reptiles, almost unchanged since the giant 4 metre long *Archelon* glided through the Cretaceous seas a

WAGER

WARD

Reef corals thrive only in clear shallow water between 25° and 30°C. In tropical Asia these conditions are found on many stretches of coast between but not near to the mouths of rivers. At low spring tides much of the coral is exposed, revealing a rich fauna of crabs, molluscs and other animals.

The leatherback turtle is the largest existing species and may grow to over 2 metres and weigh over 500 kilograms. Like all marine turtles it comes ashore to bury its eggs in the sand above the high tide mark. Turtle eggs are a delicacy to the local people and leathery turtles and other species are seriously endangered by egg collectors.

hundred million years ago. Today's largest turtle is the leatherback, a 500 kilogram heavyweight that can reach 2 metres in length. Every year the female turtles return to the breeding beaches in Ceylon and on the east coast of Trengganu, Malaya. Here, at night, the great beasts heave and snort their way over the sands. Using her powerful flippers, the female leatherback excavates a deep hole where she deposits her precious load of 150 or more white eggs, each the perfect replica of a ping pong ball. Hardly pausing to rest, she scoops the damp sand over the eggs and lumbers awkwardly back to the warm ocean; her telltale tracks betray the location of the clutch.

Smaller hawksbill and green turtles lay on tiny islands off the coasts of Borneo and Celebes but even here they are not safe. All the turtles are seriously endangered because of the predations of human egg collectors. Careful conservation laws have now been passed in an attempt to ensure that some hatchlings return to the sea each year.

Another marine creature seriously endangered by Man is the sea cow or dugong. Sightings of dugongs with their fish-like tails and twin mammae gave rise to the tales of mythical mermaids and sirens popular among ancient mariners. Dugongs are large, lethargic animals, well insulated with layers of fat and totally adapted for a simple, aquatic life. They can grow up to 3 metres long and may weigh up to 300 kilograms. They are a valuable catch as the oily meat is highly rated in the local markets. Unfortunately they are so easily caught in the long nylon nets that fishermen now use that they are rapidly heading towards extinction in the shallow seas around Ceylon, the Bay of Bengal, the Malay Archipelago and the Philippines where they were once abundant.

Dugongs can dive under water for up to ten minutes at a time and feed on sea grasses and weeds which they tear up from the sea bed, shake free of sand and devour completely. These creatures are of enormous ecological importance to Man for, in common with the related manatees of the Atlantic,

WAGER

Rhizophora zone

Sonneratia zone

Mangrove swamps reclaim land from the shallow sea. Below: The pioneer genus *Sonneratia* roots at the low tide level and can withstand almost continual submersion in salt water. Its roots bind the mud together allowing more silt to build up behind, where *Rhizophora* trees replace *Sonneratia* in a tangle of aerial roots. Further back the *Bruguiera* trees stand above the high tide mark and where the new land is completely dry, rain-forest can grow. Thick stands of nipa palms line the clogged water courses that bring down more silt through the swamp estuaries, wash-off from the inland forests.

they are able to convert the higher marine plants like seaweed into protein suitable for human consumption. Properly farmed these animals could make a useful addition to the world's food supply. If the slaughter is continued and the dugong is exterminated by overhunting a unique chance to make use of the vast resources of marine plants which at present Man cannot exploit, will have been lost for ever.

Mangrove swamps

Of all the coastal habitats, nothing supports such a variety of life as the extensive mangrove swamps. Here examples of every animal class can be found within metres of one another, all dependent on the mud which is washed down from the uplands by the great rivers and trapped among the twisting mangrove roots.

Mangrove swamps claim land from the shallow sea, close behind the ever growing coral reefs. As quickly as silt is deposited it is colonized by the mangrove seedlings. The estuaries of many of the larger rivers march out to sea fifty or more metres every year, gradually producing the great lowland plains so vital for the paddy culture of tropical Asia.

From its seaward to its landward limits the mangrove swamp can be divided into distinct zones. Next to the sea grow the primary colonizers the *Avincenna* and the *Sonneratia* trees. These have extensive underground roots which bind the mud together. The roots show great tolerance of high salinity as they are above sea level only during the lowest of low tides. Since no air can penetrate this saline mud the *Sonneratia* roots throw up vertical respiratory shoots which are exposed at low tide and allow the necessary gaseous exchange. Closer to the land is the belt of *Rhizophora* trees, the familiar tangle-rooted mangrove. Each tree is supported by a mass of branching stilt roots which, like the *Sonneratia* shoots, can respire when exposed at low tide. The *Rhizophora* trees form a virtually impenetrable barrier to the traveller and the brackish puddles, deep oozing mud and swarms of mosquitoes combine to make this one of the least attractive of places to visit. Behind the *Rhizophora* lies the *Bruguiera* belt whose roots are only covered twice a month at high tide. Again the roots loop up out of the mud to respire and the trees are supported by fin-like buttresses. Nipa palms crowd thickly in brackish water behind the mangrove line or mix among the *Bruguiera* with occasional *Pandanus* trees.

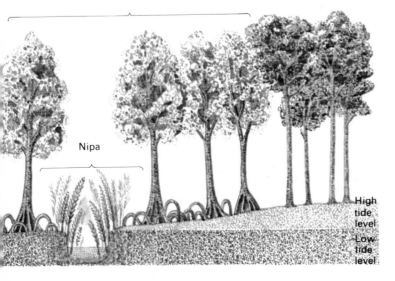

Bruguiera zone   Rain-forest

Nipa

High tide level

Low tide level

Still further back is the peat swamp forest, rich in spiny, climbing rotan palms. Above the floodline of the rivers true rain-forest grows on solid land. In some places the whole profile from sea to forest may occupy only a few metres but in south-east Sumatra the swamps may lie back 100 kilometres from the shoreline. Wherever freshwater streams wind through the swamps, fishing villages of stilt houses have sprung up. The fisherfolk fit well into the ecology of the mangrove swamp, catching fish and exploiting the other resources. Mangrove trees provide them with firewood, nipa leaves are good for thatch, *Pandanus* for baskets and matting, and rotan canes for fish traps.

At low tide the exposed mud flats are the home of armies of small crabs and fish. These fish, so amazingly independent of the water, are mud-skippers, semi-terrestrial gobies. They have stalked eyes which are regularly blinked down into an internal pocket of water to keep them moist during these land forays. Another amphibious adaptation, necessary for free movement on land or in the water, is the gill chambers which can be filled with a mixture of air and water; so long as they remain wet, the fish can breathe. When the mudskipper catches and feeds on a crab or insect the gill chambers, which are connected with the mouth, lose their moisture and the fish must hurry back to the sea to replenish its air and water supply. It has less trouble in keeping its skin wet – a quick roll in a puddle is sufficient for this.

Mudskippers can leap with a flick of the tail or even skim across the mud and water by a series of hops but normally they move slowly, crawling or 'crutching' on their stiff pectoral fins. Back in the water they revert to the normal fish mode of swimming and respiration. But the mudskipper's most spectacular locomotor feat is its ability to climb up the roots of mangrove trees. Its pelvic fins are fused to form a strong ventral sucker by means of which it can cling onto a vertical surface while it pulls itself up higher with the pectoral fins. Even on rocky shores these amphibious gobies can cling tight while heavy waves break over them.

Different species of mudskipper occur in the different zones of the mangrove swamp. The common golden-spotted mudskipper lives on the landward side of the beach but the larger blue-spotted mudskipper occupies the liquid mud at the edge of the sea. The disputes of the males of this species are fascinating. Two males confront each other, head on and mouths agape, and occasionally lock and heave to and fro. With their 'territories'

established the males build crater-like mud burrows which quickly fill to form small pools. To attract a mate to his pool to spawn, the male mudskipper gives his courtship display, erecting his colourful crest and leaping up into the air.

The climbing perch is also a mud-crawling fish. This species has evolved an air breathing lung so that it can stay out of water even longer than the mudskippers. Indeed it may even drown in water if it cannot surface to gulp air. The perch's climbing ability has been somewhat overrated. It gained its name and reputation from the fact that it was often found high up in trees but it is now known that this is the work of crows and kites which have dropped the fish there.

Another fish that frequents the mangrove swamps is the famous archer fish. At high tide these leaf-shaped, banded fish hunt among the mangrove roots for insects and grubs. Archer fish cannot survive out of water but they have evolved their own mechanism for capturing prey from above the surface; they shoot them down with a powerful jet of water, and can knock an insect off an overhanging twig up to a metre above their head. As the insect falls the fish leaps to catch it. The water spout is produced when pressure from the gills and tongue forces water from the mouth through a special groove. Having selected its target, the archer fish rises up to the surface to aim; this ensures that its line of sight is not distorted by refraction at the air/water interface.

Crowds of small active crabs scuttle over the mud flats. There are many types but most of them feed on the film of algae and plankton left by the falling tide. These crabs are never still; their front feet dance up and down carrying food to the mouth. Male fiddler crabs are rather hindered in this activity as one front leg, left or right, is enormously enlarged. This huge claw is brightly coloured and is used in displays between males and as a visual signal to attract a mate. The colour of the claw varies according to the species of fiddler crab. Orange, red, yellow or white, the large claws are effective signals when waved up and down or rotated in a circular beckoning motion which also varies from species to species.

All these crabs live in jealously-guarded burrows. As soon as the tide recedes they emerge, rolling out balls of mud and sand to clean the burrows. These are their escape holes where they run if danger threatens in the form of storks, herons, monkeys or domestic ducks. After several hours of feeding, poaching neighbours' holes, scaring off rivals and

POLUNIN/NHPA

POLUNIN/NHPA

Male fiddler crabs have one of their claws enlarged and brightly coloured. Used for signalling, the colour varies with the species of fiddler crab. The other claw, which is small and spoon-fingered, scrapes food from the surface of the mud. While feeding, their eyes are held vigilantly erect, alert for any moving object which will send them scooting for their burrows.

The little fish called mudskippers are the most conspicuous members of the mangrove fauna at low tide. Some have the ventral fins united to form an adhesive sucker which enables them to climb the smooth, wet mangrove roots. The pectoral fins are used as limbs as they move up the roots or over the mud.

courting a mate, the crabs are faced by the returning sea. They retreat into their holes and block up the entrance to keep out the water, trapping an air bubble which enables them to breathe. The little soldier crabs do not occupy permanent burrows. When the tide comes in the crab constructs an igloo of mud balls around itself, then burrows vertically downwards. It plasters the displaced sand onto the roof of its home and burrows deeper, taking its valuable air bubble with it.

One small crab *Dorippe* has a rather unusual association with another animal species, a sea anemone. *Dorippe* has its back two pairs of legs modified as little claws that carry an oval plate on which a small sea anemone lives. The arrangement is mutually beneficial. The anemone gains a firm base for attachment, a difficult achievement in the

mud flats of the mangrove swamps, and also benefits from scraps of food dropped by the crab. In return the crab gains protection, for the anemone has a nasty sting which discourages predators from attacking the combination.

The strangest of the mudshore crabs is not a real crab at all. Named the king crab it is really an ancient marine chelicerate related to the great sea scorpions that inhabited the oceans some 200 million years ago, before the reptiles. King crabs come onto the intertidal muddy beaches to breed. The female carries two or three thousand eggs on her legs, and digs a shallow depression in the mud. The male clings to her back and fertilizes the eggs as she deposits them. The eggs are then covered over. Despite their fierce armoured appearance and long, sharp tail, these ten-legged animals are quite harmless and native children run up and down the beaches catching them to strip off the tasty eggs.

Above the mud flats, in the mangrove trees themselves, live numerous weaver ants. They occur in other habitats, in the rain-forest and in cultivated fruit and rubber plantations but it is in the mangrove swamps that they are most abundant and conspicuous. Columns of red worker ants scavenge over the ground and bushes, rushing to attack anything that moves. Having found an edible morsel they drag it back up into the mangrove tree; if it is too heavy for one ant several will co-operate to carry it back to the nest. The nest is made from six or more leaves woven together with fine silken thread. It is guarded by large black soldier ants and houses the colony's strange, green queen. If the nest is damaged or torn by a predator hundreds of ants swarm out to drive off the molester and carry out repairs. Worker ants stretch between the torn edges, clinging to one side with their feet and to the other with their jaws. More and more ants join until gradually the two edges are drawn together. Other workers sew the tear with fine adhesive silk. The workers themselves bear no silk-producing organs but the larvae do. Each grub is clasped firmly in the jaws of a worker which passes it to and fro like a loom shuttle, squeezing out a fine sticky thread.

The bite of the weaver ant is extremely painful and few predators trouble these aggressive creatures. One animal which specializes on a diet of weaver ants, however, is a small spider *Amyciaea* which lives with the ants and is even tolerated inside their nests. The unusual feature about this spider is that, although it has only two body divisions compared with the ant's three, its movements and colouration mimic those of the ants so perfectly that it is not easily distinguished from them. The spider runs backwards and bears a false ant head with two very realistic 'eyes' on its abdomen. Whether it is this similarity of appearance that enables the spider to bluff its way past its

POLUNIN/NHPA

The proboscis monkey is the largest animal wholly confined to mangrove. It is allied to the langur monkeys, but the nose is grotesquely developed and in adult males is long and bulbous. Proboscis monkeys are found only around the coast of Borneo and feed almost entirely on the leaves of *Sonneratia* trees.

ATTENBOROUGH

At night a remarkable display can sometimes be seen in the landward mangrove: a whole bush alternately lit up and darkened by the synchronous flashing of hundreds of tiny points of light. The lights are produced by male fireflies that gather to flash their species signals to attract females to the group.

Left: Weaver ants repairing a rent in their nest. Weaver ants make nests in trees and bushes, fastening the leaves together with silk. The ants themselves cannot produce silk directly, but make use of their own larvae. Holding the small white grubs in their jaws, they pass them to and fro, squeezing out a filament of liquid silk.

POLUNIN/NHPA

dangerous prey is unknown, but it is certainly not easy for a human to spot the spider among the ants. When it is hungry the spider lowers itself on a silk thread, seizes an unsuspecting ant and devours it, out of reach of any of its comrades.

Where the mangrove adjoins the rain-forest, long-tailed macaque monkeys are frequent visitors to the muddy banks, feeding on crustaceans and even swimming in the estuary. A more permanent resident of the mangroves, however, is the proboscis monkey. These big, reddish primates are found only around the coast of Borneo, though here they are abundant. Proboscis monkeys live in large groups, sometimes fifty or more animals together, but they split up into smaller parties when food is scattered. They feed almost entirely on the leaves of the *Sonneratia* trees along the water front, and are more easily seen from boats than from land. They are excellent leapers and swimmers and can stay submerged for several minutes. They will swim across wide creeks to feed on estuarine islands and occasionally one has been washed out to sea and picked up by a fishing boat.

The proboscis monkey is so named because of the long, bulbous red nose of the adult male. Not only does the nose lend a deep, nasal tone to the honking contact calls of the group, it also plays a part in the threatening behaviour of the big males. Whenever the male gives his menacing bark, the long nose erects and its bright colour draws attention to the display. Females and young have neat retroussé noses.

As evening falls the doves and egrets settle to roost among the tall *Bruguiera* trees and the mosquitoes that have been abundant throughout the day become even more active. Despite their insistent hum, there is one animal still likely to draw the naturalist into the mangrove swamps by night, the *kelip-kelip* or synchronous firefly. As the sun sets the first firefly starts to flash high in a landward mangrove, and others in the same bush follow until the whole plant twinkles like a Christmas tree. Gradually the rhythm becomes quite synchronous so that each insect flashes in unison. The dimmer flashes, out of phase with the rest, are the females attracted to the males' efforts. Other flying fireflies home in on the bush until it is alive with lights. Occasionally a bat catches one as it flies and a glowing discarded tail plummets to shine for the last time on the mud below.

This is certainly one of nature's most spectacular sights but it raises two interesting questions. Why are these fireflies gregarious and why do they flash in unison? Perhaps one explanation is the presence of weaver ant colonies on a high proportion of the mangrove bushes. If a firefly landed in one of these bushes it would within seconds be carried into the weaver ant nest. Other fireflies signalling in a bush indicate that this is a safe place to land. Having thus become gregarious, they must become synchronous so that their species-specific signal pattern is clearly recognizable and not obscured in the multitude of flashing insects. Several different species of firefly occur in these swamps and all use a different flashing rhythm. A whole bush flashing the species signal must be more eye-catching than a single male. By pooling their efforts the males can probably lure females from a greater distance and the competition for a female's favours only begins when she is at close quarters.

The Monsoon Forest and Savanna

In the Indian monsoon forest the trees stand bare of leaves in summer, not in winter. During the hot, dry season (top) few of the plants display any green at all. As the wet season begins, the leaves grow again (above).

Right: The great Indian rhinoceros is the largest of the Asian rhinos. Its natural habitat is marshy grassland, and it still lives in this type of country in small areas of Nepal, Bengal and Assam. Without protection it would soon be exterminated by hunters.

Previous page:
Water buffalo have a long history of domestication and are widely used by Man in those parts of Asia where the climate is hot and wet. They are used for ploughing rice fields, for pulling carts and for milk and meat. They work best early and late in the day and like to spend the noon-time heat wallowing in mud or water.

Peninsular India, eastern Java and the Lesser Sunda Islands all have marked seasonal rainfall and in their natural state supported great tracts of deciduous woodland or monsoon forest. Not only do the trees lose their leaves during the dry winter months, these deciduous woodlands are also less tall, less luxuriant and less diverse than the evergreen rain-forest. Fewer plant species occur and in the Indian monsoon forest the *sal* trees predominate. In slightly drier areas, where the trees are more scattered and there is a good growth of ground herbage, the forest gives way to more open savanna habitat. Tree clearance by Man and his practice of regularly burning the herb layers to stimulate new growth for his grazing cattle have resulted in artificial spread of *maidans* or savanna at the expense of the monsoon forest. Originally this expansion of pastureland may have given a great boost to India's wild populations of grazing ungulates but the recent increase in domestic herds and hunting by Man have now seriously reduced the numbers of the native fauna.

Jheels or water holes are common features of the deciduous scene. They are mostly of natural origin, depressions in the landscape which fill with water during the wet season to form shallow lakes. During the dry months the pools shrink to small water holes surrounded by a muddy apron. These water holes are extremely important to the local animal life and one of the best ways of seeing the forest animals is to keep watch by a jheel where they come into the open to drink.

In the monsoon forest and savanna the whole ecological bias is rather different from the rain-forest, where the emphasis is on an abundance of small arboreal animals. Because of the limited water supply at certain times of the year, plant productivity is reduced in the deciduous habitat but those plants that do occur are more accessible to ground-living mammals. Moreover the seasonal nature of the food supply favours large species that can build up reserves to tide them over the leaner months. The result is that there are numerous large herbivores, especially in the savanna habitats.

At the height of summer the Khana Reserve in Madhya Pradesh can support 60 to 70 kilograms of wild ungulates plus 180 kilograms of domestic stock per hectare of land. A similar situation prevails in eastern Java where the Baluran Reserve supports great numbers of ungulates, mostly banteng and sambar deer. To say that a given area of savanna may at any one time support a greater mass of animal life than a comparable area of rain-forest

is rather misleading. What should be compared is the turnover of biological material. In the tropical rain-forest turnover is fast because the small arboreal creatures grow and reproduce quickly. While the savannas may support a greater standing crop of animals, these are large, slow-growing, slow-reproducing beasts with a lower rate of metabolism. The annual sum turnover of energy by the animals of the savanna is probably less than that for the residents of an equivalent area of rain-forest.

Some of the larger inhabitants of the Indian deciduous forest are animals that are more widespread in the rain-forests. Elephant, muntjac and the ubiquitous wild pig and sambar deer all occur in both habitats. The Javan and Sumatran rhinoceroses that used to occur in India are, however, now extinct there and the Indian rhinoceros is today confined to the swampy regions of Assam and Nepal. India has its own species of mousedeer, closely related to those of South-East Asia but its great wild cow, the gaur, is merely the Malay seladang under another name.

In addition, however, India has several species of deer and antelope that do not occur in the rain-forests further east. Perhaps the most attractive of

these is the chital or axis deer. These are small, elegant, spotted deer, usually found in herds of five to ten animals, though they may sometimes congregate in larger groups, particularly when grazing is good during the monsoon season. Axis deer occur in deciduous forest and savanna regions of India and Ceylon. Their requirements are simple – grass, water and shade. They are not exclusive grazers, however, and will browse on herbs and bushes and, where they can reach them, the leaves of trees.

A smaller stockier relative of the axis deer is the hog deer, which forages in smaller groups of only two to five animals and prefers rather swampier meadows than its larger cousin. It is found in the great river valleys of north India. Another deer which shows preference for moist river valley meadows is the swamp deer or barasingha. This is a large deer and the stags have fine antlers with up to twelve points. Barasingha were once abundant and travelled in herds of hundreds of animals. Today large herds can still be found near the Sarda river but elsewhere they have become extremely scarce, being found only in small groups in localized areas of suitable habitat. The barasingha is more exclusively a grazer than axis or hog deer, feeding on green or dry grass and also on some water-weeds.

Unlike deer, who drop their bony antlers regularly, the horns of antelopes and other bovids are permanent. One woodland species of Indian antelope has the unique distinction of having an extra pair of horns. The four-horned antelope is a small, pair-living animal inhabiting wooded hilly parts of the country. In contrast, chinkara gazelles live in small parties and prefer more open habitat. Two larger antelopes, the nilgai and blackbuck, also range into the open glades of deciduous woodlands, but they are also inhabitants of the semi-desert and will be considered with the desert fauna.

Wild water buffalo still occur in a few parts of northern India. They are longer-horned than the domestic and feral water buffalo that are so abundant throughout the Oriental region. Some of the so-called wild populations of water buffalo in other parts of India are merely feral animals which have escaped from domestic stock. As their name implies, water buffalo prefer to live in swampy grazing areas and in consequence they are ideally pre-adapted for working conditions in the waterlogged paddy fields of tropical Asia. They have a long history of domestication and have spread with Man through South-East Asia to the Greater Sunda Islands.

In drier areas the forest gives way to open savanna country, with scattered trees and grasses.

Axis or chital deer are found in deciduous forest in India and Ceylon, where there is grass, water and shade. Usually found in small herds of five to ten animals, they are the only species of deer in which both sexes have a spotted coat at all ages. Their dappled pattern blends well with the light and shade of the monsoon forest.

The four-horned antelope is a relative of the larger nilgai and its two pairs of horns make it unique among ungulate mammals. It is usually seen in pairs, not herds, in wooded hilly country.

Following page: A herd of elephants comes down to drink, the cool water in front of them, the dry dust blowing up behind. The elephant's daily water requirement is large and they never wander more than a day's journey from a water hole or a river. Unlike their African relatives, cow elephants in Asia do not have tusks and in Ceylon many of the bulls are also tuskless.

The leopard has a very wide range over Africa and the warmer parts of Asia, but its numbers are now everywhere reduced by hunting. In dry grassy country its spotted coat serves to conceal it and so enables it to stalk its prey and capture it with a rapid rush or pounce.

The tiger lives in a very wide range of habitats, from the cold coniferous forests of Siberia to the hot jungles of south and south-eastern Asia. It has a greater fondness for water than any other large cat and often swims or wades in rivers to drink or to cool its body.

Domesticated buffalo are used as beasts of burden to draw ploughs, pull heavy carts or tow felled logs out of the forest. They require no confinement and never wander far from the owner's village, returning each day to wallow in the mud baths that are left for their use.

The three main predators of the Indian woodlands are the wild dog or dhole (which has already been discussed in its rain-forest environment), the tiger and the leopard. Indian leopards belong to the same species as those that inhabit the rainforest but they are generally spotted rather than the melanic or 'black' panther form of South-East Asia. Despite their differences in colour, leopards and panthers have almost identical habits whether they are in rain-forest or deciduous woodland. Although occasionally active by day, they usually hunt at night, taking wild pigs, deer and monkeys, which are found in both habitats.

The tiger is probably India's most famous animal though now, unfortunately, it is becoming very scarce. Tigers are rather solitary animals. Males wander over many square kilometres, their ranges overlapping those of several females. Mating takes place during brief consortships and the females look after their cubs alone, depositing them in a den under a rock overhang or in a cave. They hunt primarily by night, taking the larger species of deer, occasional pigs and sometimes gaur. Where natural prey have become scarce they turn their attention to domestic stock and even humans; there are many authenticated cases of man-eating tigers. It is not surprising therefore that tigers are regarded as poor neighbours by the local villagers.

In India jackals tend to live around human habitations, and their eerie nocturnal howling is characteristic of the Indian rural scene. They live by scavenging, but are not wholly beneficial to Man as they will take poultry and young or sickly sheep and goats.

Top: The striped hyena is a powerful animal with immensely strong jaws and huge molar teeth. These are probably adaptations for scavenging, for they can crush the largest bones of an ox or buffalo carcase to extract the marrow. The hyena also has an ugly reputation as a robber of shallow, ill-protected graves.

Right: Cheetahs were once common in India and survived well into the present century; the last wild cheetah was shot in 1947. They were formerly used for hunting blackbuck but as they were not bred in captivity, but caught and tamed as adults, this probably led to their decline and extinction.

Persecution by Man has caused a great decline in the numbers of tigers. Only an estimated 2,000 now survive in India compared with 40,000 in 1930 – and their long-term survival will only be possible within large, well-protected game parks.

The fate of India's big savanna cats, the cheetah and lion, is even sadder: the cheetah has become extinct and the lion population has been reduced to less than 300 animals. The cheetah was once a common sight hunting on the plains and its disappearance seems to be due to the scarcity of its chief prey, the blackbuck and chinkara. The age-old princely sport of hunting with trained cheetahs may well have accelerated the animal's decline for since cheetahs will rarely breed in captivity the packs had to be replenished with newly caught animals. Eventually the wild population reached

land and savanna. Tigers, leopards and lions here occupy a central position in the food chain, preying on the herbivores and providing food for the vultures, hyenas and other scavengers that rely on the carcases they discard. The importance of the cats to their prey is enormous: they cull out the weaker members of the herds and help regulate the total numbers of ungulates so that overgrazing and degradation of pasture is avoided.

Another carnivore of African origin, the striped hyena, is found on the plains. Hyenas are not bold, savage killers like the great cats. They will kill young and sick animals, but they make their living by scavenging off the carcases left by other predators. They will even dig up buried corpses unless these are covered with heavy stones. Hyenas are powerful animals with immensely strong jaws and molar teeth which enable them to crush the bones of even the largest carcase to feast on the rich marrow.

Another scavenger/hunter is the jackal. These small dogs can usually be found lurking around a lion or tiger kill waiting for the bigger predators to move away so that they can rush in to rescue the leftovers. Jackals are, however, quite able to hunt for themselves and catch lizards and small mammals. They also serve a useful social function, cleaning up much of the debris of human towns and villages. Their plaintive calls are characteristic of the Indian night.

Vultures, too, arrive as if by magic whenever there is a kill in open country. Wheeling high in the sky, discernible only as a small black speck, the vulture scans the plains with its sharp eyes. When it picks out a carcase far below it plummets down to the ground. Other vultures take note of this descent and move across to investigate. Quickly a crowd of the birds assembles. Some habitually fly so high that they cannot possibly see what is happening on the ground but from their vantage point they can keep watch on numerous other vultures and make sure that none will feed alone.

The first birds to spot a dead or dying animal are the white-backed vultures but they do not take precedence at feeding. The 'king vulture' of India is the black vulture, the only species with a beak big and sharp enough to rip through the hide of a large animal. They are solitary birds and usually only one or two arrive at any carcase. The white-backed and the long-beaked vultures sit around jostling, ready to move in when the black vultures have had their fill. They fight and squabble, hissing and screeching at one another and dancing about

too low a level to support this regular cropping. The last wild Indian cheetah was shot in 1947.

The Asiatic lion is slightly heavier in build than its African relatives. The young are less spotted, the adults have bigger tail tufts and the males have shorter manes. They are, however, quite obviously members of the same species. The lion was once widespread throughout Arabia and northern India. In the last century one 'sportsman' shot over 300 of them. By the early 1900s, however, they had become confined to the Gir forest in north-west India. A recent attempt has been made to reintroduce a few of the Gir lions into another reserve in the state of Uttar Pradesh but it is still too early to say if this has been a success.

The disappearance of the big cats has many serious consequences for the ecology of the wood-

PETER UGANDER/N

with outstretched wings in their battles for precedence. At the bottom of the hierarchy are the little Egyptian vultures whose small size and narrow pointed beaks are more suited to picking out the guts and offal than to tearing off the meat. The head and neck of all these vultures are naked, as a covering of feathers would be continually soiled by their bloody probing.

One bird which is dominant to all the vultures at a kill is the adjutant stork whose enormous pointed beak is respected by all. Like the vultures, and for the same reason, the adjutant stork is naked to the shoulders and also bears an ugly bald pouch under the throat. Although traditionally water birds, these scavengers can be found in dry regions far from water. When not associating with vultures they sometimes hunt for frogs and locusts.

Two smaller bird scavengers are the black kite and the house crow. Both have developed an association with man and stay around human habitations, taking advantage of the plentiful scraps.

In the closed habitat of the rain-forest, birds of prey are uncommon but in the Indian woodlands and savanna they abound. These open spaces are ideal hunting grounds for raptors that rely on good

A king vulture, distinguished by its red head and neck takes precedence at the kill over white-backed vultures and the smaller particoloured Egyptian vultures. The king vulture's beak is sharp and strong enough to rip through the hide, while the Egyptian vultures have narrow pointed beaks for picking out guts and offal. A pariah dog takes the place of a scavenging jackal.

The savanna food chain is simpler than that of the forest. Photosynthetic plants trap solar light energy to produce primary foods – leaves, flowers and fruits. Animals feed on these, either directly as herbivores or indirectly as carnivores or scavengers.

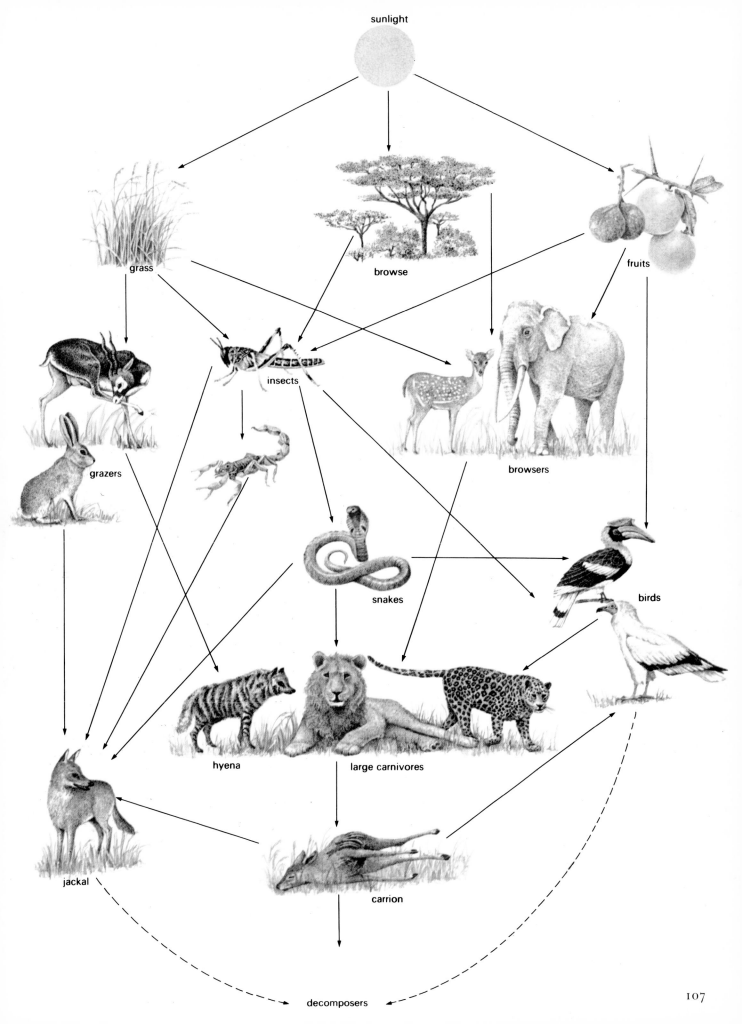

sunlight

grass

browse

fruits

insects

grazers

browsers

snakes

birds

hyena

large carnivores

jackal

carrion

decomposers

107

vision for detecting their prey. Great eagles circle high over the plains and mountains. The tawny eagle sometimes joins the vulture at a kill but most eagles hunt their own prey. The mountain hawk eagle and Bonnelli's eagle take small mammals and reptiles while the grey-headed fishing eagle lives over water, snatching fish from under the surface. Peregrines, laggers and red-headed falcons prey on smaller birds, catching their victims by sheer speed. The Shikra sparrowhawk scores because of its great ability to twist and turn through the obstacles of the woodlands. The small black-shouldered kite hovers like a kestrel, watching for insects, rodents and lizards or rests in a tree or on convenient telegraph wires.

At night the owls take over, each specializing in its own type of prey. The big forest eagle owls hunt for birds; barn and grass owls take small mammals; the brown fish owl takes fish and the spotted owlet insects.

The insect-eating birds in India include tailor-birds, drongos and bee-eaters related to those found in the rain-forests of South-East Asia. Sun-birds also occur in both areas, feeding on insects and nectar. Two bird species which are more specifically inhabitants of the open country are the hoopoe and the roller. Both are big, colourful birds feeding on large insects, lizards and the like. The roller perches on a dead tree and scans the ground for prey, while the hoopoe hunts by walking and running on the ground. Both nest in natural holes in trees.

Indian babblers are also ground feeders. There are several species, all group-living. They are unusual in that members defend communal terri-tories and only one pair of the group will breed. All the other birds help to incubate the eggs and feed the young. Possibly this habit enables them to survive in scrub areas where a single pair would be hard-pressed to rear a brood alone. While the rest of the group are feeding on the ground one bird always sits in a nearby bush acting as sentinel. It can give two calls: one of alarm if danger threatens and another to indicate it is hungry, when a replacement will fly up to take over the watch.

The seed-eating weaver birds also live in groups but each pair makes its own intricate nest in a crowded tree colony. The baya weaver is particu-larly common in open country. The females are dull-coloured but the males in breeding plumage are resplendent in bright yellow and black. The nests are built of grass, woven tightly together to form a hollow flask-shaped structure, rounded at

MILLER

Though the Indian python does not grow as long as the reticulated python (its maximum length is around 6 metres) of the two it seems to prefer bigger prey. The chital deer fawn being swallowed whole will be assimilated by a prodigious feat of digestion. A meal like this will last the python for several weeks.

Above, left: The Indian roller is a common bird all over the lowlands in open forest, around villages and in gardens. Perched on a bare branch – or a telegraph pole – it watches the ground for the large insects and small reptiles on which it feeds. It gets its name from its habit of rolling and somersaulting in flight.

The baya weaver looks rather like a sparrow, but the male has a yellow crown and black throat in the breeding season. The remarkable flask-shaped nest is made of ribbon-like strips torn from blades of grass, with mud in the lower part to give it stability. Most of the nest building is done by the male.

the bottom and tapering up to a point where it is attached to a branch. At one side, near the base, a narrow tube entrance hangs down. Mud is incorporated in the bottom of the nest to reduce its sway in the wind.

Weaver nests on terminal twigs are fairly safe but those near the trunk sometimes attract the unwelcome attention of snakes. India has many snakes and they take a heavy annual toll in human lives. Although cobras are often blamed for these fatalities the real culprits are usually the nocturnal kraits and Russell's viper. Only one species of true cobra occurs in India though regional forms differ in their markings. Thus the spectacled cobra of southern India and Ceylon has a pair of rings on its expanded hood while further north cobras have only a single ring in the middle of the head. Hindus revere the cobra as a good snake and regard it as a symbol of fertility. They feed and worship cobras at great animal festivals. In non-Hindu areas, however, cobras are badly treated, being killed for their decorative skins, pitted against the skill of tame mongooses or kept in poor conditions by India's famous snake charmers. Snake charming is an ancient ruse, for cobras have no ear membranes and cannot hear the musician's flute. The 'charmer' relies on the normal behaviour of cobras which is to rise cautiously, swaying gently, when the top of their dark basket is removed. There is little danger involved, as most of these performing snakes have had their fangs broken off.

Grey and small mongooses are often thought to be immune to snake venom. This is only partly true. Mongooses are slightly less susceptible to the poison than other animals, but a full bite by a well-stocked cobra would certainly be fatal. The mongoose employs delaying tactics, prolonging the duel and baiting the cobra to strike. It evades each blow until eventually the snake has lost so much venom and strength that the mongoose can dash swiftly in for the kill. In the wild mongooses rarely tackle snakes but make an easier living off rodents, lizards and frogs.

The Indian rock python is another common snake. It is smaller than the reticulated pythons of the rain-forest but at six metres is quite formidable and will tackle large prey. One instance has been reported of a rock python devouring a child.

Other Indian reptiles include skinks, large monitor lizards and colourful agamids. Of the arboreal agamids the *Calotes* lizard is perhaps the most common, an active little creature with a big head, long tail and a row of frilly spines along the back. It is brown or yellow and the males become bright red on the throat and head when fighting or displaying during the breeding season. The *Calotes* lizard is quite harmless but its colourful flush makes it look dangerous and has given it the name 'blood-sucker'.

Rather surprisingly one of the most serious enemies of Indian snakes and lizards is the gorgeous blue peafowl. (The green peafowl is found in Burma, Thailand and Java.) These beautiful birds have been domesticated for at least 2,000 years but are still abundant in the wild, frequenting woodland meadows and the open clearings around natural water holes. Apart from reptiles they take insects and vegetable food. The male peafowl or peacock bears long, fan-like tail coverts decorated with iridescent eyes which it displays during courtship. Each morning and evening the cocks emit raucous screeches to announce their whereabouts. During the day these sharp-sighted birds are always vigilant and they are often the first to raise the alarm when a leopard or tiger approaches. Although they spend most of their time strutting on the ground, peafowl are powerful fliers and roost, like other pheasants, in the safety of trees at night.

Apart from the obvious relationships of predator and prey there are other more indirect benefits that may be gained from association with another species. Parasitic animals, like ticks and lice, feed on many vertebrates but at a level that the host can afford. Dung beetles work busily on the unwanted waste of the large mammals, rolling away neat little balls which they bury as food for their grubs. Starlings and cattle egrets accompany the grazing herds, catching the insects that are disturbed by the multitude of moving feet.

Some of the rodents are quite capable of living independently and do so in the woodlands but they prefer to exploit Man whenever possible and infest his dwellings. House rats are a serious menace for not only do they foul human homes and food, they also carry a multitude of fleas which are the vectors of bubonic plague. Larger and less harmful are the common bandicoot rats which may reach up to 80 centimetres in total length.

The largest of all the Asian rodents is the Indian porcupine. Porcupines like rocky hillsides but adapt themselves to any kind of country. They occupy deep burrows several metres long, which usually terminate in a large nesting chamber. Their burrows can be recognized by the great quantity of ejected earth and bones that lie scattered around the entrance. The bones are not the remains of a

The small Indian mongoose is often found near human habitations and is easily tamed. Mongooses do good service by destroying rats and mice in houses and grain stores, though they will take poultry and their eggs as well. Their habit of killing snakes has been much exaggerated.

Top: Cattle egrets have become close associates of Man in rural districts of Asia. They also gain advantages from their association with cattle or buffaloes. From a perch on the grazing animal's back, they fly down to catch grasshoppers and other insects disturbed by the trampling feet.

Cobras rear up and spread their hood to warn enemies to keep away. The venomous bite will not immediately incapacitate a man or a large animal, and the snake has no interest in biting anything but small creatures which it can swallow whole. The pattern on the hood varies throughout its range, the 'spectacle' mark being characteristic of Indian cobras.

Green peafowl are found further east than the more famous blue variety, in Burma, Thailand and Java. Both species eat many different types of food, from insects, grain, seeds and fruit to snakes and lizards.

The feathers that make up the blue peacock's great ocelated fan are those of the lower back, the upper tail coverts, and are raised by the tail itself. During its impressive courtship display the peacock quivers its spread feathers so that the iridescent colours catch the light from different angles. Although they spend most of their time on the ground, peafowl fly well and roost in trees at night, out of the way of ground predators.

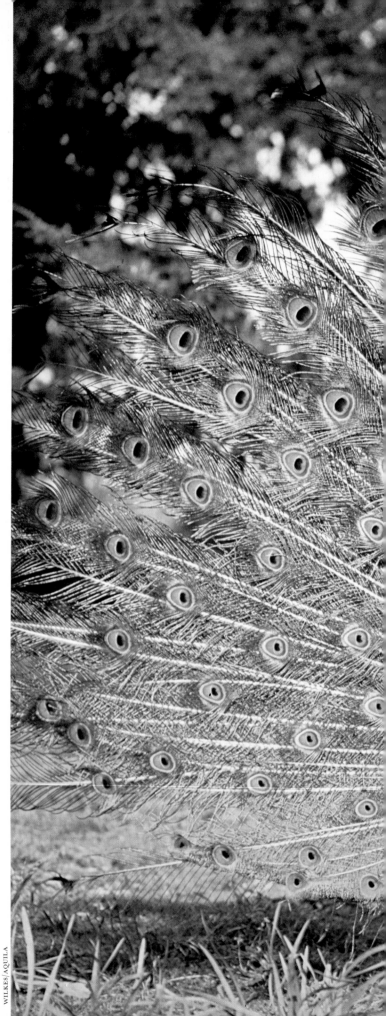

porcupine kill, however. Porcupines collect them during their nightly wanderings and gnaw them to obtain much needed calcium. Otherwise their diet is strictly vegetarian.

The hog badger and ferret badger are also burrowers and creatures of the night. Both have boldly patterned, black and white faces which presumably act as a recognition signal for other badgers they may meet in the dark. The markings may also serve as a warning to would-be predators that they have a powerful bite and sharp claws. Unlike the vegetarian porcupines, badgers are omnivorous and eat a good deal of animal material, eggs, worms and insects, as well as fruit and shoots.

Monkeys and squirrels

Several kinds of monkeys live in India. Once these were all forest or woodland species but they have been able to adapt their way of life to cleared land and some have even become quite urbanized. This has only been possible because of the extreme tolerance shown to monkeys by the majority of the human population. To the Hindus langur monkeys represent the god Hanuman and are therefore sacred. Several species of langur are found in the Indian subcontinent; the nilgari langur is found in the hilly forests of the south-west and the purple-faced langur occurs in the damper parts of Ceylon. Commonest of all is the grey langur.

The grey langur is found throughout India and Ceylon but shows a good deal of local variation in colour, hair pattern and tail posture. Grey langurs live in large troops, usually of twenty or so animals but up to a hundred in some places. They are mainly leaf-eaters but accept bark, buds and fruit when these are available. Each group occupies a habitual range which varies in size according to the richness of the habitat. In dry regions group ranges may be several square kilometres in extent but in optimal conditions they can be as little as ten hectares. Breeding is strictly seasonal and is timed so that the infants are born in the spring, coincident with the arrival of the summer monsoon and the richest flush of new plant growth.

In addition to their unusual relationship with the local human population, grey langurs also have a curious association with axis deer. These two species are frequently found close together and several authors have suggested that this is a mutually beneficial arrangement: that the monkeys are alerted to danger, perhaps in the form of a marauding leopard, by the alarm bark of the axis deer or,

conversely, that the deer may be warned by the cries of the monkeys. It would be surprising if wild animals did not recognize the significance of the alarm calls of other common species but such behaviour does not necessarily indicate that it is solely for reasons of safety that deer and monkeys associate. Moreover, lengthy observation of these monkeys revealed no tendency for them to seek out or to stay with deer. It is far more likely that the deer follow the monkeys whenever these are feeding in trees that the deer enjoy as browse but cannot normally reach. Langurs are rather messy feeders and drop a good many twigs, leaves and fruit which are quickly devoured by the deer. Axis deer have even been observed browsing on the leaves of a branch which, weighed down with monkeys, had swung within their reach.

By eating mature leaves, the only food available in the dry season, langurs are able to survive in some areas where no free water is available for several months. This is not possible for the macaque monkeys, who prefer younger shoots and rarely eat mature leaves. Their distribution is determined by the year-round availability of drinking water. Two common species of macaque occur in India, the rhesus monkey north of the Godavari river and the bonnet macaque to the south. A third species, closely related to the bonnet macaque, is the toque macaque of Ceylon.

Since they enjoy a mixed diet and are always ready to experiment, macaques have become quite urbanized and in some areas rely almost entirely on the human population for food – either donated or stolen. They frequently live in temples: there are great opportunities for obtaining food from visiting worshippers and the grounds are often planted with their favourite natural foods, strangling figs such as the banyan and pipal, and tamarind and mango trees. Temple monkeys have developed quite different habits from their forest counterparts; their home ranges are much more restricted and the animals are generally more aggressive among themselves. Even forest monkeys, however, will raid human crops and villages whenever the opportunity arises. It seems that macaques have lived in such close proximity to Man for so many thousands of years that this commensal relationship must now be regarded as a natural condition for them.

The bonnet macaque, like the rhesus, lives in large groups with a strict dominance hierarchy among the males. There is a slight excess of males in these groups, which is unusual among primates;

RAO/NHPA

The lion-tailed macaque lives only in the mountain forest of western India. Unlike the other Indian monkeys it is shy and seldom seen, and has failed to adapt to Man's environmental changes. The lion-like tuft of hair at the end of its tail has given it its name.

The common langur or Hanuman monkey is really a forest animal, but is established in towns and villages in most parts of India, where it is regarded as sacred. As they are safe from molestation, these populations have no fear of man, and they do considerable damage to gardens and growing crops.

GOODERS/ARDEA

an excess of females is more common. Whereas in most macaque species subadult males are attacked by their superiors and driven out, in bonnet macaques there is a greater inter-male tolerance and young males stay with the troop. The explanation for this probably lies in the behaviour of bonnet macaques when they meet a predator. While the rest of the group rush for the trees, the males drop back to form the first line of defence against the attacker. The higher number of males may provide a greater margin of safety under stress of predation.

Unfortunately one species of monkey in India, the lion-tailed macaque, has failed to adapt to Man's environmental changes. Black, with a long mane of grey hair and a tuft of hair at the end of its short tail, it is totally unlike any other macaque or, indeed, any other monkey. Little is known of its habits and there is now serious concern for its survival. It is estimated that less than a thousand lion-tailed macaques survive in isolated pockets of forest in the Western Ghats.

Another traditionally arboreal group of mammals, the squirrels, is also well represented in India. The Malabar squirrel and the rock squirrel are related to the giant squirrels of the rain-forest. Perhaps the most familiar squirrel of the Indian peninsula is the five-striped plantain squirrel. It is abundant in parks and town gardens as well as in the forest. These squirrels travel about in small parties of five or six. They are active, noisy creatures with a shrill bird-like call and eat fruits, buds and bark, with occasional insects and birds eggs.

Water holes

Elephant, leopard, deer, gaur and monkeys all come regularly to water holes to drink. In addition to providing refreshment for the land mammals these oases also support a rich fauna of their own. There are fish, snails and leeches but most conspicuous of all are the water birds that congregate, often in vast numbers, on India's jheels and artificial reservoirs. Some are only visitors, for many ducks migrate to India each winter from temperate regions further north. These include such familiar European species as shoveler, teal, garganey and pintail and also greylag geese. Residents, like the spot-billed duck, also frequent the water holes. Some of these native ducks are seasonal in their behaviour. About six thousand lesser whistling teal visit one lake in Calcutta every day during the winter months but never in summer. Then they remain on the swampy feeding grounds where the

RAO/NHPA

reeds are long enough to afford them cover and shelter.

One of the most fascinating aspects of the water birds is their great diversity of beak form. The ducks all possess broad flat bills designed for sifting and sorting through surface matter on shallow water. Slight variations in beak form correspond to differences in feeding habits. The most aberrant beak of all is found in the shoveler which uses its heavy spatulate bill to skim through the shallow water and muddy ooze. Bristle-like teeth act as a sieve to strain out the small organisms on which it feeds.

The beaks of the cranes are strong, pointed and highly functional. Cranes probe the soft ground of their marshy feeding areas in search of insects, worms and frogs as well as tender shoots and berries. The pretty demoiselle crane and the rarer Siberian white crane visit India only in winter but the tall sarus cranes breed there regularly. Like other cranes, they indulge in strange hopping, leaping dances during their annual courtship. They show great boldness in defence of the piles of rushes and reeds that form their nests and, since pairs often mate for life, they are revered as a symbol of conjugal devotion by the country people.

The heron family have rather long, sharp beaks,

GOODERS/ARDEA

Shoveler and pintail (distinguished by the white neck-stripe) are among the great variety of ducks that frequent the shallow lakes and reservoirs of the Indian plains. Many are winter visitors which breed far to the north and are familiar species in temperate Asia and Europe.

RAO/NHPA

Far left: The Malabar or Indian giant squirrel has a wide range in India and its handsome black, brown and cream pattern varies regionally. Nearly a metre in overall length, it lives among trees, making large nests in which it sleeps at night and for much of the day as well.

The five-striped palm squirrel is a lively, noisy little animal, usually seen running and climbing in groups of half a dozen or so, constantly uttering its shrill bird-like call. It is abundant in towns and villages in India, wherever there are parks and gardens with growing trees.

ideal for spearing fish. Grey, purple, night and pond herons all breed at Indian water holes, together with the related egrets. The egrets, and particularly the cattle egret, show how easily a beak for catching fish can be adapted for catching insects. Several kingfishers have also adapted their traditionally fish-catching beaks to a successful insect-hunting way of life.

Storks are larger birds with long, powerful beaks suitable for probing among tall reeds for insects, frogs and lizards. Resident painted storks and migrant white storks use their beaks with all the skill of a Chinese using chopsticks. The heavy adjutant stork has a more powerful bill, which can tear meat off a carcase and frighten off jackals and vultures. The open-billed stork is something of a specialist, feeding almost exclusively on big water snails which have thick shells and a tough plug or operculum at the entrance. Its long mandibles can only close at the tip and it uses the open middle of its bill like a pair of nutcrackers to break open the heavy shell and extract the soft body of the snail.

The darter has a heron-like pointed beak for although it chases fish underwater it can draw back its long neck and strike whilst swimming fast behind its prey. Cormorants also swim after shoals

Like most herons, cattle egrets nest gregariously in trees. When seen around villages they are usually pure white but in the breeding season the head and neck become orange and orange plumes develop on the back and breast.

The tall sarus crane stands one and a half metres high and is a resident bird in the Indian lowlands. Its nest is a mound of reed and rush built on marshy ground, and the birds are bold in its defence. However, they are regarded with affection by the country people and are seldom molested by Man.

The painted stork is a common bird in well watered districts throughout India where it is often seen, alone or in small groups, wading in the shallows in search of frogs, fish and freshwater crustaceans.

of fish but have broader, hooked bills with which they grab their prey. They then surface to juggle the fish around before swallowing it head first.

Related to both the cormorants and the darters are the spotted billed and rosy pelicans. Both have long, hooked beaks and large skin pouches suspended from the lower mandible. Pelicans feed in a unique manner, hunting in groups and driving fish towards the shallows. Each bird ducks synchronously with a sideways sweep of the bill, using the pouch like a dip net to catch the fish. They hold the fish in the pouch until the water drains out before swallowing it.

Pelicans breed in only a few scattered parts of India although they are widespread for most of the year. They nest in trees in crowded colonies with storks and other birds. The young are fed on partly

The open-billed stork gets its name from its most obvious feature: the closed mandibles meet only at the tip, leaving a gap between them for most of their length. As the bird feeds mainly on large water snails this is probably an adaptation for extracting them from their shells, but its exact working is still not understood.

In its breeding plumage the pheasant-tailed jacana is a beautiful bird, but the long, down-curved tail is lost at the end of the nesting season. Jacanas spend most of their life on floating plant leaves. Their toes and claws are greatly elongated, an adaptation which enables them to walk on water-weed or waterlily leaves, and their nest is built on floating vegetation.

digested fish which is regurgitated. Chicks can sometimes be seen with their head and neck deep inside the parent's wide gape, probing for food. Pelicans seem fairly clumsy and many youngsters fall out of their nests. As the parents pay no further attention to these unfortunates, only those that are nearly fledged survive.

Another variant on bill shape is found among the ibises. Both the white and glossy ibis have long, slender, down-curved beaks which are used to sort through the mud of the lake bottoms in search of crustaceans and worms and also for catching insects on land. A more elaborate bill has been developed by the related spoonbill which also sifts the mud for food. Spoonbills wade slowly forwards sweeping their broad, spatulate bills from side to side. The mandibles are held slightly open and the lower one rakes up the mud, together with the tadpoles, insects and crustaceans that it eats.

Other waders also have long bills for probing the mud. The stilt with its long, spindly legs and long beak can feed in this manner several metres from the shore, whilst, at the other extreme, the common snipe forages on the fringe of marshy pools. The painted snipe feeds in the same way but is remarkable for quite different reasons. These birds are common around the edges of jheels where there is plenty of vegetation. They exhibit an unusual instance of inverted sexual behaviour. Instead of the male bird wearing bright plumage and courting a female, as is usually the case amongst birds, the male painted snipe is dowdily coloured. It is the larger female who initiates courtship and who is beautifully 'painted'. Moreover, as soon as she has performed her inevitable duty of egg laying, she leaves the male in charge to incubate the clutch and rear the young.

Perhaps one of the most familiar birds of Indian water holes is the blue-breasted banded rail that creeps shyly among the reeds seeking insects and vegetable food. The related gallinules are bolder and venture out onto the floating weeds. Both the purple gallinule and the common moorhen have rather unspecialized beaks for they are omnivorous and search through the surface vegetation for anything edible, either plant or animal. Their elongated feet enable them to walk calmly over the flimsiest of flotsam. The more aquatic coot has developed flaps of skin on the side of each toe, a sort of webbing which greatly improves its swimming ability.

Elongation of the toes is taken to extremes by the jacanas or lily trotters who spend their lives almost entirely on floating plant leaves. They tread daintily over the water hyacinths and lilies in search of insects, molluscs and occasional unwary fish. Bronze-winged and pheasant-tailed jacanas are both widespread in India and nest on the floating plants among which they feed. Even though their eggs are sometimes partly submerged on a water-logged leaf they seem to remain viable.

Of all India's water birds none are as picturesque as the huge flocks of greater flamingoes that nest on the lakes of the Great Rann salt flats near the Pakistan border and also in Ceylon. Not only are these gorgeous pink birds eye-catching on the ground and in the air but they also have perhaps the most bizarre beaks of all water birds. Flamingoes feed with the upper bill underneath the lower mandible which, together with the tongue, helps to pump water and disturbed mud in and out of the mouth. Coarse particles are barred by stiff excluder hairs. Small diatoms, algae, molluscs and crustaceans are sieved free by fine filtering lamellae inside the mouth, worked onto the tongue and then swallowed. This extremely complex filtering structure is essential since the soda and salty water of the lakes flamingoes inhabit would be toxic if swallowed in large quantities. In fact, the flamingo is able to sieve its food almost dry and so avoid ingesting the solutions of detrimental salts.

Flamingoes are gregarious birds both when feeding and nesting. The nests are curious mud domes which project above the shallow water and have a concave depression at the top. Here a single egg is laid and hatched by the ridiculously crouching parent, whose folded legs protrude to the rear. After a few days the grey fluffy youngsters leave the nest and are herded together with a few adult attendants who act as 'baby-sitters' while the parents feed elsewhere. Young flamingoes have straight unspecialized beaks and are entirely dependent on their parents for food until, by the third month, their own filtering system has developed.

After the summer monsoon the water-level of the jheels begins to drop. The marshy perimeters become rock hard and the shallow pools shrink to small water holes ringed by bare earth ploughed by the hooves of thirsty deer and cattle. Where the water vanishes completely from the surface, elephants may dig deep holes to find it below. With breeding over, the great congregation of water birds disperse over wider feeding areas to visit the more permanent lakes and reservoirs, rivers, swamps and paddy fields until the next monsoon signals nesting time again.

The spotted-billed pelican is the common pelican of India. They are seen mostly in open water, where parties fish efficiently by cooperative action, driving the fish before them and scooping them up in the shallows. They nest in trees, often in company with painted storks.

Top: Spoonbills and white ibis breeding in company. The birds of the heron family generally nest in colonies in trees and quite often several species will nest together. Both spoonbills and ibises have specialized bills: the ibis uses its long, down-curved beak to sort through mud in lake bottoms as well as for catching insects. The well-named spoonbill also feeds in shallow water, raking up tadpoles, insects and crustaceans with its lower mandible.

Camels are efficient conservers of water and have an unusually high tolerance of dehydration. When they do drink, however, they can take up to a third of their body weight in ten minutes. Special adaptations of the red blood cells ensure that sudden dilution of the blood has no ill effects.

The Desert Wastes

Desert ecosystems are found in areas that receive less than 25 centimetres of rainfall a year. Most of the great deserts of Asia lie in the Palaearctic region but in the north-west of the Indian subcontinent, in Sind and Rajasthan just east of the Indus valley, lies the Thar or Indian Desert. This area was not always desert; in fact 2,000 years ago it was covered by jungle but Man's poor agricultural practices, timber-cutting activities and over-grazing by his stock have turned the region into an arid plain where few wild animals can survive. This process has accelerated within the last century so that the desert is currently extending around its perimeter by eight kilometres every decade.

Apart from the perpetual snow and ice of the poles and the highest mountains, desert is the environment most hostile to life. Two main problems face the desert animals – great heat by day and scarcity or lack of water. Air temperature at midday often rises to 45°C and may exceed 55°C while the ground surface is even hotter, reaching 70°C or more. At night the ground loses heat and can become very cold. Differences in air temperature of 30°C between day and night are not unusual. Rain is so scanty and spasmodic that it cannot be relied on as a source of water. In some years there may be no rainfall at all.

Animals in this hostile environment can combat the heat problem in one of three ways: they can tolerate it, they can evade it or they can keep cool by evaporation of water or sweating. The last method is very expensive in a habitat where water is not readily available.

Only the insects seem to be able to tolerate great heat. It is not unusual to see insects flying in the heat of midday or a scarab beetle pushing its little ball of dirt over the hot surface of the sand. Insects are very successful at living in dry places and possess several adaptations that help them conserve water. Their outer cuticle is almost impermeable to water and the respiratory system is made up of fine tubes where only the finest, terminal ends permit water to escape. Some insects can even extract water from the air.

All other animals fall into the last two categories. As a general rule approximately two-thirds of the body weight of any animal consists of water. They lose water in several ways: by evaporation from the surface of the skin, from the moist respiratory surfaces during breathing, with the urine and faeces and, in mammals, with the milk they feed to their young. Yet if it is to survive an animal must remain in water balance. Water lost must be offset by water taken in. In the desert where water intake is limited the only animals that can survive are those that reduce output and even those that combat desert heat by evasion show special adaptations for conserving vital moisture.

Several species of scorpions are adapted to arid environments. Like insects they have a hard cuticle or exoskeleton which prevents water loss from the body. Additional protection against overheating is afforded by their habit of 'stilting', raising themselves off the hot ground to allow the circulation of air beneath the body. In spite of these adaptations against desiccation, they hide in rock crevices and under stones during the heat of the day and emerge only at night to hunt. Scorpions are entirely carnivorous and feed mainly on insects; they catch and hold their prey with their large pincers and paralyse and kill it with venom from the tail.

Desert vertebrates show remarkable physiological adaptations to conserve water. They have efficient kidneys which secrete highly concentrated urine and specialized intestines that remove water from the faecal material before it is deposited. Where possible they also reduce the water lost by evaporation, for this is the most significant factor in water balance. Small mammals have no sweat glands. Nor do reptiles though snakes and lizards still lose some water through their scaly skins and during respiration, for expired air is saturated with water vapour.

Of the four living orders of reptiles, three – the snakes, lizards and tortoises – are well represented in deserts. They are all poikilotherms or 'cold-blooded', so that their body temperature fluctuates with that of their surroundings. In the early morning when the air temperature is low the desert lizards – agamids, lacertids and a variety of geckos – use the heat of the sun's rays to warm their bodies. When warm they avoid further exposure by seeking out shade or burrowing under the sand. Most desert lizards are diurnal and can move swiftly over the hot ground to catch insects and avoid hawks. Many have toes fringed with small, elongated scales which provide a 'snowshoe' effect when the animal runs over loose sand. Snakes are more nocturnal and avoid the day-time heat. The saw-scaled viper hunts at night, seeking lizards and small mammals, especially gerbils, whose burrows it invades.

Reptiles from arid and desert areas excrete most of their nitrogenous waste as uric acid in the form of a thick white paste so that they lose little water in this way. Lizards are also able to secrete salts through the nose and this enables them to feed on

ROSS

ROSS

KINNS/AFA

The spurges (*Euphorbia*) resemble the cactuses of the New World and like them are mainly plants of dry regions, adapted to conserve water. Most spurges have acrid, poisonous sap which helps to protect them against browsing animals. Insects which are adapted to feed on them acquire their ill-tasting properties and so are protected against birds and other predators.

Top: Sparse vegetation in northern India. Overgrazing by goats and other domestic animals, together with uncontrolled felling of trees, has made a desert of what was flourishing forest two thousand years ago, and the process still continues.

All scorpions are carnivorous, catching their prey – here a young locust – in their claws and killing it with venom from their tail-borne sting. Many species of scorpions inhabit deserts; they hide in the heat of the day and emerge at night to hunt any small animal they can catch and overpower.

those plants with a high mineral content that occur in arid regions. But the main physiological advantage of desert reptiles is their poikilothermic nature. This means that they have a high energy turnover when active, but that this decreases when they retire underground. Consequently the resting reptile respires more slowly and loses less water with expired air. Moreover as its body temperature falls to approach that of its surroundings it loses less water by evaporation, especially if the cool retreat is relatively humid.

Small mammals, though 'warm-blooded', react to the heat in much the same way as the reptiles. They have no defence against overheating since to save water they do not sweat or pant. Because of their high surface area in relation to their weight they would require large quantities of water (which is unavailable) to dissipate the heat and their only recourse is to avoid it. They excavate burrows and either venture out only at night or, if diurnal in habit, during the cooler hours of the early morning and dusk. The extreme temperature variations measured at the surface do not persist to any great depth and a burrow one metre underground enjoys a fairly constant temperature of about 30°C.

The Indian desert gerbils are burrow-dwellers. They are pale grey, mouse-like rodents, with long, tufted tails and live together in small colonies in the Thar desert and surrounding semidesert areas. Their faeces are dry and their urine very concentrated and, although they do not normally encounter it, they can drink sea water without ill effect, so used are their kidneys to dealing with high concentrations of salts. Perhaps their most extraordinary adaptation, however, is their ability to live on perfectly dry food with no water at all. All their body fluid is derived from metabolic water. As organic food, such as carbohydrate, is assimilated and oxidized to provide energy for the body, it breaks down to form water and carbon dioxide. The carbon dioxide and much of the water vapour is expelled during respiration but sufficient water is retained to satisfy the gerbil's needs.

Another inhabitant of the Thar desert is the long-eared hedgehog. Although it does not live in burrows it shelters in shady places during the day and is active mainly at night when it feeds on a diet of insects, lizards and birds' eggs. Its large ears probably serve as efficient heat radiators like those of the American jack rabbit. When it is resting in the shade of a shrub or rock where the ground temperature is relatively low it is able to lose heat by radiation to its cooler surroundings. The

BROEKHUYSEN

Several species of Agamid lizards live in hot, dry deserts. As they are cold-blooded, their temperature varies with that of their surroundings. When they become too hot in the sun they burrow into the cooler ground or seek shade under rocks and stones.

enormous ears, which are well supplied with blood vessels, greatly extend the animal's exposed surface area and facilitate this heat exchange.

The large mammals

Only large animals can afford to combat heat by evaporation or sweating. The camel is probably the most remarkable and best known of all desert animals. Although the two-humped Bactrian camel exists in the wild state only in the cold deserts of Tibet, the domesticated camel is used for transport throughout the hot, dry regions of Asia. There is

The Asiatic wild ass is a very different animal from the ancestor of the donkey, whose home is North Africa. It ranges from Mongolia and Tibet to Iran and the southern U.S.S.R., with a small population in north-western India. Its numbers are everywhere much reduced by hunting.

The Arabian camel no longer exists in the wild state, but is clearly adapted to live in hot deserts with very little available water. It is now a valuable servant of Man in very dry regions and is used both for riding and as a pack animal.

one record of a camel caravan having travelled 944 kilometres between two water holes. Admittedly this was in the winter and early spring, the coolest part of the year after scattered rainfall had produced fresh vegetation for moisture and feed, but even in the driest months camels are capable of quite remarkable feats.

How do they manage? Contrary to popular belief they do not store water in either the hump or stomach. They do, however, store fat in the hump and this is a very efficient way of carrying food reserves. When the fat is oxidized to provide energy metabolic water is produced, as with the gerbil, though this is of little help since the camel loses more water during ventilation of the lungs than is gained in this way.

Several adaptations contribute to the camel's success in the desert. First, its size. Although the camel must be in danger of overheating, it is too big to escape the midday sun by seeking out shelter as smaller animals do. Its larger size, however, gives it an advantage over the smaller animals of a much lower surface area in relation to its mass. This means that it heats up and cools down more slowly. Its thick, insulating coat also ensures that temperature changes affect it only gradually.

The camel's second adaptation to desert life is its heat tolerance, which also enables it to conserve water. Water that is used to cool the body through perspiration is irretrievably lost but the camel manages to conserve water by not sweating until its body temperature reaches 41 °C. (Man in the desert begins to sweat at 37 °C to prevent overheating.) Not only does the camel require less water to maintain a body temperature that is much nearer the temperature of its surroundings, but such heat tolerance means that the camel can go for several more hours before it needs to sweat at all. During the cold desert night it loses the excess heat without drawing on its precious water reserves. The camel's thick coat and tolerance of temperature changes is again of great advantage for temperatures can fall to as little as 34 °C by morning.

As well as being an efficient conserver of water, the camel also has an unusually high tolerance of dehydration. Most mammals losing 20 per cent of their body water would die but the camel can lose over 40 per cent without any obvious ill effect. The reason for this is that the camel does not suffer from a thickening of the blood, as other animals do when deprived of water, but loses water from its tissues. The albumen content of the camel's blood is very high and this raises the osmotic pressure of the blood

so that it can retain its water even when the surrounding tissues are seriously depleted. Another important adaptation of the blood allows the camel to drink without restraint even when it is dehydrated. When it does fill up, a thirsty camel can take 100 litres of water, up to a third of its body weight, in ten minutes. In Man and most mammals the flow of water into the blood caused by excessive drinking after dehydration results in rapid swelling and rupture of the red blood cells, often with fatal effects. The camel's red cells, however, are very resistant to sudden dilution of the blood and it comes to no harm.

A second large desert mammal, the Asiatic wild ass, is becoming increasingly rare but a small population still lives to the south-west of the Thar Desert, on the salt flats of the Little Rann of Kutch. During the monsoon this region may be covered by half a metre or so of water, partly from the floodings of local rivers and partly with water blown from the sea by the strong south-west winds. During the dry months, November to June, there is no water and the islands of the wet season become small hillocks or *bets* supporting some grass and a few trees, similar to the desert scrub of the mainland. Throughout this semidesert the wild asses

GRANDJEAN

Small, mouse-like gerbils are a group of mammals highly characteristic of deserts. They live in burrows in the sand, subsisting on dry seeds, but even on this dry diet, their remarkable physiological adaptations enable them to exist without drinking any water.

The blackbuck once ranged widely in the plains of India and the deciduous forests, but destruction of forest cover and merciless hunting has made it rare. Now mainly an inhabitant of semi-desert country, it is one of the few large mammals that is found only in India.

roam. They are beautiful animals with pale, sandy-coloured coats and dark chestnut manes and dorsal stripes. Their lower parts are white and the shoulders, saddles and sides of the rump are fawn. This colouring is disruptive and was no doubt evolved as a protection against their natural predators, wolves, which are now extremely rare. Asses, like the camel, have a great tolerance of dehydration and can withstand a water loss of 30 per cent of their body weight. The ass restores its water content to a normal level by drinking even faster than the camel; this may also be a protection against predators lurking near water holes.

Adjacent to the Thar Desert lies a belt of arid tropical thorn forest, a vegetation type which is widespread throughout the plains of Rajasthan, west Punjab and Gujarat and in the shadow of the Western Ghats. These areas may receive more than 75 centimetres of rainfall a year but temperatures are high, reaching a maximum of 49°C, and water is scarce. A few scattered small trees – acacia, mimosa and cassia – rise above the understorey of thorny, xerophytic shrubs and the sparse cover of grasses. Excessive use of the forest by Man for firewood and grazing his livestock is turning this semi-desert into desert, capable of supporting even less life.

The isolated semidesert of west Rajasthan is the main stronghold of the blackbuck, once probably the most abundant wild hoofed animal in India. The blackbuck is an animal of open terrain, a species of the thorn and dry, deciduous forest. With the destruction of forest cover it has adapted to the wastelands and cultivated areas, where it is shot because of the damage it does to crops. Blackbuck are well adapted to life on the plains. Unless the temperature in the shade is more than 32°C, they rarely seek the shade of trees but remain in the open meadows. They are able to tolerate direct sun better than other wild ungulates and are more diurnal in their grazing and foraging. Even in the hottest months they can survive without drinking. Like other hoofed animals that live in arid environments on forage of low food value, the blackbuck can probably recycle nitrogen within its body instead of excreting it in urine. This not only conserves water, it enables the animals to survive on food with a low protein content. Most fawns are born shortly before the rains so that they will benefit from the fresh, new vegetation that springs up after the monsoons.

Nilgai or blue bulls also occur in the same thorn and dry deciduous forest. They are horse-sized antelopes, the males resplendent in coats of grey, black and white with conspicuous white throat patches and the females and calves a less obvious, tawny colour. Whereas blackbuck are almost exclusively grazers, nilgai take some browse and will even stand on their hindlegs to feed from the upper branches of thorny shrubs which other animals cannot reach.

These antelopes were once preyed upon by the Asiatic lion and the elegant Indian cheetah. Today, however, the cheetah has become extinct and the lion extremely rare, restricted to a mere 1,300 square kilometres in the Gir forest on the Kathiawar Peninsula. Other carnivores occur in the true desert, seeking out burrows and shelters wherever possible to avoid the heat. Both cats and dogs are able to withstand high air temperatures, greater than their body temperature, when humidity is low. Both species lose heat by panting and the dog is an especially good heat regulator. Desert carnivores are able to subsist without drinking water because they obtain sufficient fluid from the bodies of their prey. In the Thar desert the desert cat is common, preying on birds and small mammals. Its range extends westward into North Africa – it is the cat that was domesticated by the ancient Egyptians. A subspecies of the red fox, known as the desert fox, also inhabits the dry regions of north-west India, and preys on small rodents and birds.

Desert birds

Birds have a considerable advantage over the other desert vertebrates – the ability to fly. Whereas a small rodent may have a range of only a few hundred metres, a bird can fly over considerable distances in search of water. Sandgrouse, for instance, are never out of reach of water and will travel 50 kilometres or more to water holes where they congregate in huge flocks in the early morning. Moreover if conditions become too unfavourable within their environment birds can move away, either to some other site within the desert where local rainfall has improved the availability of food and water or to a completely new habitat. Although several birds do pass most of their lives in deserts they show behavioural adaptations for coping with the extreme habitat rather than physiological adaptations as in the mammals.

According to their size, desert birds select different microclimates in which to spend most of the day. Most desert birds are small and conse-

Nilgai, like blackbuck, are found only in India and live in the same thorn and dry deciduous forest. They feed both on ground vegetation and on the leaves of thorny shrubs that other smaller animals cannot reach.

Birds are far more mobile than mammals and can fly long distances to drink. Sandgrouse are characteristic of deserts and can live in very dry areas. They feed on hard seeds and habitually gather in large numbers at lakes for water, which they carry to their young in their crops. Although quite unlike them in appearance and habits, they are allied to pigeons.

quently suffer from the same problems of heat regulation as small rodents. They avoid the extreme heat of midday by feeding in the early morning and late evening and roosting in the shade during the hottest hours. Larger birds, however, remain active, soaring on thermals at great heights where the air is cooler.

Reproduction raises some problems peculiar to birds. Desert species nest on the ground and the eggs and nestlings must be shielded against over-heating. Most birds, like the desert lark, nest in a depression in the shade of a rock. Once the young birds have hatched the parents have another problem, supplying their brood with water. For a long time it was believed that breeding sand-grouse soaked their breast feathers with water during their morning drinking sessions and that the young sucked the feathers when the parent returned to the nest. Unfortunately this attractive theory did not explain how the water escaped evaporation during the long, hot flight home and it is now known that the chicks receive water by the more mundane method of regurgitation.

Desert larks are distributed widely throughout the dry regions of southern Asia and are almost all coloured to harmonize with their surroundings, ranging from almost black in deserts of black lava to palest, sandy brown. Similar cryptic colouring, a defence against predatory hawks, foxes and desert cats, is also found in the see-see partridge, the only truly desert game bird of India.

The great Indian bustard is a much larger bird which inhabits the dry open country. It can fly quite well but prefers to run on the ground. It has specialized three-toed, blunt-clawed, flat feet, well suited to this stony terrain. The bustard is omni-vorous, eating seeds as well as insects and even small vertebrates when it can catch them. Un-fortunately these spectacular birds have been quite unable to withstand human pressure and the species has become seriously endangered. Bustards make large tasty meals and although they are shy of men on foot, they are easily approached by hunters on camels or in jeeps. It is ironic that the bustard, like the cheetah and lion, is threatened when its habitat, the desert scrub, becomes more extensive every year.

In early historic times lions were abundant in south-west Asia and in the lands of the eastern Mediterranean but they have been hunted to extinction throughout almost the whole of this region. The Gir forest in north-west India is the lion's one remaining haunt outside Africa. In this region of scrub and dry forest a small remnant survives under government protection.

Ascending the Himalayan foothills, evergreen pine forest gives way to bamboo and rhododendron. At around 3,500 metres the moorland and alpine zone begins, with, higher still, the inhospitable regions of perpetual ice and snow.

WALLER/ARDEA

The Great Mountains

To the north the Oriental region is bounded by a great mountain chain. The Pamirs, Hindu Kush and Himalayas make up the southern wall of the enormous mountain system of eastern central Asia and include the highest peaks in the world – Mount Everest is 9,500 metres. North of the Himalayas is the great plateau of Tibet and further north still, well within the Palaearctic, are more ranges, the Kunlun Shan, Tien Shan and Altai mountains. The Himalayan massif is a comparatively recent phenomenon on the geological time scale and in the past was not the barrier to animal movements that it is today. Consequently much of the high altitude fauna of this area extends over the whole complex of mountain ranges and is typical of both the Oriental and Palaearctic regions. However, in general, we shall confine ourselves to the animals of the Pamir-Himalayan chain and the Tibetan plateau.

The natural monsoon forest extends well into the foothills of the Himalayas so that on the lower slopes the tiger, antelopes and wild pigs that are typical of northern India are still found. But the foothills also support some animals found nowhere else in the region. Until very recently it was thought that the hispid hare and pygmy hog had become extinct; it was only in 1971 that these two species were rediscovered in the Manas Sanctuary in Assam. The hispid hare has shaggy, bristly fur and is about the same size as a common hare, but its ears are shorter, giving it a rabbit-like appearance. It is found in the metre high thatch that covers the drier grasslands. The pygmy hog is a creature of both the jungle and thatchlands. Blackish-brown suffused with red, an adult stands a mere 23 centimetres at the shoulder. Pygmy hogs move around in small herds of fifteen to twenty animals. Males are fearless, attacking, charging and cutting at intruders. The groups can move at lightning speed with the animals staying close together and piling on top of one another when they reach a new refuge of thatch.

The fauna of the foothills and the ranges beyond are not well documented and there are still expeditions here to search for such mythical creatures as the yeti and the buru. As recently as the 1950s E. P. Gee was able to establish the existence of a 'new' species of monkey, the golden langur. Golden langurs are a golden chestnut colour in winter, paler during the hot season; the young are practically white. They move about in troops ranging in number from seven to thirty, feeding in trees and descending to the ground for water and salt.

Higher up the mountains the environment becomes more extreme, the vegetation changes and there are fewer animal species. In Nepal for instance, over a distance of little more than 160 kilometres, the biomes range from tropical evergreen forest in the south, through temperate broadleaved oak and coniferous forests to the alpine associations of the high mountains and the treeless steppes beyond. In addition to the influence of altitude changes on ecology there are marked east–west variations due to the decrease in monsoon rainfall towards the west.

The bamboo and rhododendron forest

With increasing altitude the evergreen pine forests give way to thick bamboo and rhododendron forest. Although the bamboo thickets may support animals which also occur in montane forest, they contain certain characteristic species. Bamboo rats with their thick, heavy bodies, short legs and brief, scantily-haired tails are common residents here. Their stout, orange incisors are very prominent and together with their strong, digging claws are effective tools for excavating burrows beneath stands of bamboo. Bamboo rats spend much time underground, cutting and eating bamboo roots but they occasionally leave their burrows to feed above the surface on grasses and seeds.

By far the most famous resident of the bamboo forest is the giant panda or *bei-shung*, a favourite attraction at western zoos. Giant pandas are known to roam among the steep ridges and ravines in high, cold country in the Chinese province of

The golden langur is the most recent species of monkey to be discovered. It was found in 1955 by E. P. Gee and has been named *Presbytis geei* after him. It inhabits a limited area on the banks of the Bramaputra river in Assam. Adults are a striking golden chestnut, the young almost white.

Main vegetation zones at increasing altitude in the Himalayas, with some of the animals that live in these habitats. Southern facing slopes enjoy a milder climate than those facing north.

GRANDJEAN

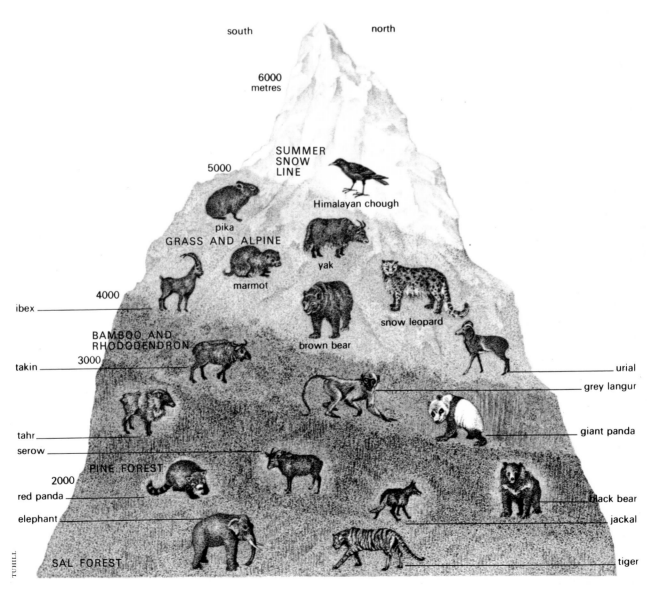

south   north

6000
metres

5000

SUMMER
SNOW
LINE

Himalayan chough

pika

GRASS AND ALPINE

yak

marmot

4000

ibex

snow leopard

brown bear

BAMBOO AND
RHODODENDRON

3000

takin

urial

grey langur

tahr

giant panda

serow

PINE FOREST

2000

red panda

black bear

elephant

jackal

SAL FOREST

tiger

TUHILL

Szechuan and, if this area were not so inaccessible, both geographically and politically, they might prove to be even more widespread. Although they have been known to take gentian flowers, fish, mouse-hares· and small rodents, they feed mainly on bamboo and have two particular adaptations to the forepaws and teeth to enable them to cope with this diet. Each forepaw has an elongated wrist bone, covered with a fleshy pad, which acts as a thumb opposed against the fingers and provides a kind of forceps with which to handle and hold bamboo stems. The huge molar teeth are blunt and studded with tubercles and ridges for crushing tough bamboo culms.

In winter the giant panda lives in a world of snow and ice for which it is well equipped with a thick coat and coarse hair on the soles of its feet. It is difficult to say why its fur bears such a contrasting pattern. It may be disruptive and so provide camouflage but it seems more likely that it would be extremely conspicuous against a snowy background. Giant pandas usually travel singly and the striking markings may well serve as a recognition signal over considerable distances. Alternatively, it has been suggested that the black and white coat may be warning coloration in the same way as the skunk's. Giant pandas are shy, slow-moving animals, incapable of beating a hasty retreat, but if cornered they can defend themselves well with their very powerful, heavy jaws. Possible predators include the snow leopard and wild dogs and although there is no evidence that they do tackle an adult giant panda, they are known to take young.

The affinities of the giant panda are hard to define. The smaller red panda is probably its closest relative and beyond this it resembles in various respects both bears and racoons. There is constant zoological controversy over which group it belongs to.

The extraordinary habitat of the giant panda is the home of a number of other little known animals including the snub-nosed or snow monkey, a big, robust langur with a long, silky mane of golden hair and a snub nose which has been described as resembling a bright blue butterfly sitting on its face.

The cat-sized, red panda is handsomely coloured, chestnut above, black below and on the legs, with white markings on the face. It lives in bamboo forest in the eastern Himalayas and seems to prefer a higher altitude and colder climate than the giant panda. It is a nocturnal creature, sleeping by day curled up in a tree and feeding at night on the

CHINESE PEOPLES REPUBLIC

ground on bamboo sprouts, grass, roots, fruits and eggs and perhaps even birds or mice. Red pandas travel in pairs or family groups. Two to four young are born in a single litter in a hollow tree or rock crevice. They are blind at birth and remain dependent on the mother until she produces her next litter.

The alpine meadows

Bamboo and rhododendron forest gives way to the higher moorland and alpine zones at about 3,500 metres. Because of the low temperatures and short summer, trees cannot live above this altitude. The dramatic change in environment on crossing the tree line is reflected by the number of animal species that cannot exist above this level. The red

The giant panda's black and white colouring may be disruptive, providing camouflage in snowy country, or it may have a social recognition function. Left: Giant pandas have only recently been observed and photographed in the wild in the Chinese province of Szechuan, where they live in the high, cold bamboo forest. In spite of their heavy build they readily climb trees. Bamboo is their staple food; their forepaws are specially modified for holding it and their huge molar teeth can crush the hardest stems.

The red panda is an animal of the eastern Himalayan and west Chinese forest. It is quite a small animal, 4 to 5 kilograms in weight, though its very thick fur makes it look bigger. Red pandas live on vegetable food, feeding at night and spending the day sleeping high in the trees.

Yak are the domestic cattle of the people of Tibet and other nearby high altitude regions. Wild yak still exist: they are black with long black horns and a little white on the muzzle. Domestic yak are smaller with shorter horns and are usually particoloured, white and brown.

Right: Wild goats live at quite low altitudes in south-west Asia and the north-west corner of India. They are the ancestors of the ubiquitous and destructive domestic goat, which has played such a large part in converting huge areas of fertile land to desert. These are feral goats, wild animals of domestic ancestry.

panda, Himalayan black bear, leopard, wild dog and langur are found in the upper reaches of the montane forest but no further. Other animals such as the markhor, musk deer and Himalayan weasel may wander above the tree line during the warmer summer months but return to lower elevations for the rest of the year.

The upper slopes of the great mountains provide a bleak and inhospitable habitat and animals living in the alpine zone have to contend with several problems, some of which are similar to those facing the inhabitants of that other extreme habitat, the desert. In both deserts and alpine zones animals have to cope with extreme temperatures. Both habitats have a shortage of available water – although the Himalayan ranges receive plenty of rainfall much of the water is frozen as ice and snow and is inaccessible throughout much of the year. Both have a short flowering season, after the brief rainfall in deserts and during the summer thaw on the mountains. In both habitats animals have problems of shelter; both are subject to strong winds that bury the land beneath sand or snow. In addition, animals living at high altitudes have to overcome the physiological problem of reduced atmospheric pressure and fall in oxygen tension.

The activity of the muscles used in breathing tends to be reduced and the animal suffers from lack of oxygen; men with 'mountain sickness' suffer from nausea and fainting. After a time, however, the body becomes acclimatized and the number of red blood corpuscles (oxygen transporters) increases, as does the acidity of the blood.

As well as the seasonal pattern of warmer summers and bitter, cold winters the Himalayas show great fluctuations in diurnal temperature so that regions above the tree line are subjected to warmth during the day and sub-zero temperatures at night. On the southern slopes of the Himalayas at a height of 4,200 metres the diurnal range in June is from about 8°C to −7°C. At high altitudes the sun's rays lose less heat to the air as they pass through the thinner atmosphere. Insolation at ground level is very intense so that the temperature near the ground is much higher than the surrounding air temperature and it is this microclimate near the ground that allows plants and the smaller animals to survive in alpine zones.

The larger mammals that live above the forest can withstand the low night temperatures without the need for cover, though they all have thick, shaggy coats for protection against the cold. The

bovid family is particularly well represented in the Himalayas. More species of wild goats and sheep live in this region than anywhere else. The largest bovid, the yak, has been domesticated for centuries by the people of Tibet but wild yak can still be found. They are bulky, black animals with a little white on the muzzle and long, black horns. In summer yak range up to 6,000 metres, feeding on the wiry grass and small shrubs that grow at these altitudes.

Four species of wild goats occur in the Himalayas. On arid, rocky hills at quite low altitudes in Sind and Baluchistan is the common wild goat, which is more widely distributed through the Palaearctic region. It is the wild ancestor of the domestic goat but a far finer animal with heavy, keeled horns a metre or more long. The highest coniferous forests are roamed by the tahr, a typical goat with rather short horns close together on top of the head. Tahrs are gregarious, grazing in herds while sentinels mount a lookout for possible danger. Even higher, on the rocky slopes and cliffs above the tree line, lives the markhor, perhaps the finest of all wild goats with its long horns shaped like thick, heavy corkscrews. Its fur is rich, reddish brown turning to grey in winter and, like all the

wild goats, it is an amazingly agile climber. In winter markhors retreat to lower slopes where forage is more plentiful. They will even climb along the branches of evergreen oaks to browse on leaves.

The tahr and markhor are true Himalayan species but the ibex has a much wider range, inhabiting mountainous regions all over central Asia from the Himalayas to the Altai and extending westwards as far as Spain. These animals live mainly above the tree line, grazing between rocky crags and on the meadows just below the snow line. They roam much higher than the smaller wild goats. The relationship of size to surface area is important in heat conservation for animals living in cold regions. Warm-blooded vertebrates, mammals and birds, can be fairly independent of their environment if they can maintain their body temperature. Heat can be conserved by reducing the body surface either by development of a compact form or by a relative increase in size. Thus the higher living ibex with its short legs and thick neck is both a bigger and more compact animal than the wild goat.

The argali is the largest species of wild sheep and extends from Tibet to Siberia. Fifteen races or subspecies have been named. Marco Polo saw and described one of these in the thirteenth century; now named Marco Polo's sheep it is renowned for its magnificent horns, which curve round in more than a complete circle. Argali live on the high plateau of Tibet, ascending to 4,500 metres in the summer and wintering in the more sheltered lower valleys. Their habitat is harsh, cold and dry and they subsist on scattered tufts of grass – the only vegetation available.

The bharal or blue sheep occurs in Tibet, Sikkim and Nepal and shows characters intermediate between the sheep and the goats. Bharals are found at altitudes of 3,600 to 5,000 metres on the rich and abundant grass of the alpine meadows. They feed in herds during the summer months, except for the old males which forage at higher elevations. During the rest of the year the ewes congregate in small harems round the males. Their brownish grey pelage with a touch of slaty blue serves as protective coloration. When they stand motionless they are hardly visible against the rocky outcrops. As an added defence sentinels are posted to keep watch over the grazing herd.

The shapu or urial is the smallest of the wild sheep of eastern Asia and inhabits grassy mountain slopes, usually below the tree line. The urial has probably contributed to the stock of domestic sheep. None of the wild sheep has the woolly fleece of the domestic breeds; instead they are covered with thick, coarse hair. Wool seems to be a feature that has arisen and been selected for during domestication.

Three kinds of 'goat antelopes' occur in the Himalayas and are relatives of the European chamois. The serow is a goat-like animal with a black coat, large ears, white beard and a greyish mane. It lives in the Himalayas in forest around 2,000 metres and also extends into the tropics in Malaya and Sumatra where it occurs at lower altitudes on steep, limestone outcrops. The smaller goral is found throughout the Himalayan range, living on rugged grassy hillsides and on rocky ground near forests. It is remarkably agile on the precipitous cliffs and screes. Gorals move about in small, family groups though the old males prefer to travel alone for much of the year. They are active during the cooler parts of the day, in the early morning and late evening. After the morning feed they drink and retire to sunny ledges to rest until evening.

Takins live in the rhododendron and bamboo forest near the timber line in 'giant panda' country. In summer large herds graze above the timber line

but in winter the animals break up into smaller groups and migrate to grassy valleys lower down the mountain. If danger threatens they dash for safety into thick undergrowth. During the day they rest up in thickets but emerge in the evening to feed, and use regular paths to their grazing areas and salt licks.

Two kinds of antelopes, the chiru and the goa or Tibetan gazelle, complete the list of Himalayan and Tibetan bovids; both occur on the high, cold Tibetan plateau. The chiru has a swollen muzzle which may be an adaptation for breathing cold and rarefied air. Chiru run in large herds on the Tibetan plains and individuals scoop out hollows in the sand in which to rest and shelter from the wind.

All these animals are adapted for life in cold, harsh conditions. Their fur is thick in the winter and thinner in the summer and the seasonal coats are often differently coloured so that they merge with the background and camouflage the animals from such potential predators as wolves and snow leopards. Apart from the two antelopes they are all rock climbers, marvellously sharp-eyed and sure-footed. Those that live at high altitudes are able to subsist on sparse, dry vegetation.

Not only the bovids benefit from the rich new growth on the alpine meadows during the summer months. The hangul or Kashmir stag moves out of the forest of birch and blue pine to graze among the higher pastures. Shortly before leaving for their summer haunts in March or April, the stags shed their antlers. The Kashmir stag is the local representative of the European red deer which were once kept as royal game for the maharajahs. The shou or Sikkim stag is another race of red deer, rather larger

Above, right: The markhor is the largest and finest of all wild goats and is a true Himalayan species. Its horns are quite unlike those of any other animal; they are like thick, heavy corkscrews and may be a metre or more long. When hunting for trophies was a fashionable sport markhor horns and heads were in great demand.

The goral, smallest of the so-called 'goat antelopes' lives in the great mountains of Asia on rocky hillsides where grass and trees grow freely. When threatened it takes to precipitous cliffs, which it climbs with amazing agility.

than the hangul, which occurs in the Chumbi valley in Tibet. Apart from natural predators the red deer suffer considerably from the activities of Man, both as a poacher and a keeper of cows, sheep and goats, which are allowed to overgraze, and so destroy the alpine meadows. Musk deer, too, are persecuted all along the Himalayas, in this case for their musk pods which are collected from the males and used in the manufacture of perfume.

Many smaller mammals live above the tree line. They, too, have dense furry coats for protection against the cold. In the high mountains of Kashmir lives the little known woolly flying squirrel, a cumbersome creature capable of gliding like its forest relatives. It frequents inhospitable rocky terrain, living on a diet of moss and perhaps lichens. Because of the nature of its habitat it does not need the sharp, clinging claws of the tree squirrels; instead its claws are blunt.

One way in which animals survive in the harsh alpine climate is by creating their own warmer environment within the habitat. This is exactly what many of the rodents and lagomorphs (the rabbit and hare family) do. They may use natural shelters and holes which protect them from the cold winds but many dig their own deep burrows. The temperature in a burrow a metre or so beneath the ground is much higher than the prevailing air temperature at night and within their own warmer microclimate the animals are able to ignore the cold outside. Père David's voles inhabit any easily tunnelled banks and slopes in the mountain meadows and have even been found as high as 5,700 metres. They are active by day and night, bustling along their surface runways. They make good use of the short summer to collect the stems and leaves of herbaceous plants, drying them in the sun and storing them away for the long bitter winter.

Two other high alpine mammals, the marmot and the pika, are both burrowers and both are common up to 5,000 metres. The Himalayan or bobak marmot is a stoutly built animal with a short, bushy tail and very small ears. The compact form and reduced extremities are important adaptations for heat conservation. Marmots live in large colonies and no doubt this gregarious habit helps them to survive at high altitudes. They feed on grass around their burrows and any individual that sees or scents danger utters a loud scream that sends the whole colony rushing underground. In autumn they drag dry grass into the burrows and in this warm bedding they hibernate until spring when the young are born. The long-tailed

marmot which lives lower down the mountain has similar habits.

Pikas or mouse-hares live in the mountains at altitudes from 2,500 metres to well above 4,000 metres. Pikas are rather similar in appearance to guinea-pigs but belong to the rabbit and hare family, the lagomorphs. Himalayan pikas live on open, rocky ground and hide under rocks or loose stones, or, if there are trees, they burrow at the roots. During the winter their homes are buried under deep snow and they are believed to live on stores of dried grass and other vegetation.

Birds, too, make use of any natural shelters they can find and some, like the Tibetan desert chat and Hume's ground jay, use the burrows of pikas to nest in. Few birds, however, can exist at great heights. Himalayan and Tibetan snow cocks, game birds like large partridges, both breed above 5,000 metres and the snow partridge feeds and nests up to 6,000 metres but all other vertebrates at this altitude appear to be visitors only. These include choughs, griffon vultures, ravens and lammergeiers, scavengers that follow man, wandering yak and wild sheep wherever they roam. Alpine choughs have been recorded visiting mountaineers' camps at over 8,000 metres. Lammergeiers are a spectacular sight, using their 3 metre wing-span to soar on the thermals.

Though there are few species in the alpine zone, a food chain is still apparent. The Himalayan golden eagle preys on marmots, snow cocks and snow partridge and the little Tibetan weasel ranges high to feed on birds' eggs. In any assemblage of animals the predators tend to be far fewer, both in number and diversity, than their prey, and

Lammergeiers are also known as bearded vultures and though they are not closely related to vultures they subsist in the same way, by scavenging carrion. They have a wide range in the Old World, but are never found far away from mountains. They are widely but quite falsely accused of carrying off lambs.

The golden eagle lives in the mountains of Asia or Europe, where it is one of the largest birds of prey. In the Himalayas it ranges up to 5,000 metres, preying on marmots and snow partridges. It also feeds on carrion, especially in winter when food is hard to find.

this is especially true of the alpine zone where the range of food is much reduced. Yet four large and several small carnivores inhabit the high rocky terrain and the upper edges of the mountain forest.

The snow leopard is a beautiful animal, with long thick fur of smoky grey with black rosettes. Its colouring is perfect camouflage against the rocky background. Rather smaller than the average panther, the snow leopard is mainly solitary, hunting at night. It preys on wild sheep, goats and musk deer and follows them up and down the mountain slopes on their seasonal migrations. Smaller victims include hares, rodents, monal pheasants and chukor partridges. During the winter months snow leopards descend to the lower valleys where they find domestic stock an easy prey. Here, too, they meet their only serious enemy – Man. They are hunted

Snow leopards are the only large cats found at great heights in the mountains. They are extremely shy and their mountain haunts (right) are often inaccessible. Snow leopards prey on wild sheep and goats, following them up and down the mountains on their seasonal migrations. They also take smaller animals – rodents and hares – and birds. Their only serious enemy is Man, who persecutes them for their luxuriant fur.

The Himalayan black bear is a forest dweller and is not often seen above the tree line. It is a good climber and usually spends the winter in a hollow tree several metres above the ground. The young are born in a den of this kind and, like those of other bears, are quite tiny at birth, no larger than rats.

for their luxuriant coats and despite recent bans on export of snow leopard pelts the trade persists.

Another large carnivore, the Himalayan black bear is a forest dweller, ascending to the tree line at about 3,500 metres in summer. It is a medium sized bear with a wide white V-mark on its chest. Like most bears it is almost omnivorous, feeding on fruit, ripening corn, wild sheep and goats and the young of the Kashmir stag. Higher up is the less predatory brown bear and to the north and east the Tibetan blue bear. The hide and scalp of a 'yeti' collected by an expedition to Tibet in 1960–61 later proved to be portions of the skin of a blue bear and a serow respectively.

Of the small predators Pallas's cat, found in Tibet, is as big as a domestic cat. Its pale grey coat with rather sparse darker markings enable it to merge with its background, a useful advantage when creeping up on prey. Its very long fur, especially on the underside of the body, protects it against the bitter cold. Pallas's cat has a broad head with short ears and eyes set high in the face. The behaviour of a specimen in captivity suggested that this enables the animal to peer over a rock without flattening its ears and so reducing their efficiency, while exposing little of the head's tell-tale outline to the watchful marmots, pikas and birds on which it preys.

High altitude invertebrates

At altitudes above 4,000 metres much of the alpine zone is covered with glacial scree and heavy frost restricts plant colonization. These stony habitats are colonized mainly by invertebrates. In this type of situation, though animals gain heat rapidly during the day, they also lose it rapidly at night. There are, however, pronounced differences between atmospheric conditions and the immediate surroundings of invertebrates, which are micro-climates on or in soil, under stones, on rock surfaces, in crevices, underground cavities and ice crevasses and under snow or vegetation mats. Within these microhabitats the animals are subjected to much less extreme temperatures. More-over as the temperature falls rapidly at night condensation occurs on rock and leaf surfaces so that water is available.

High altitude invertebrates are dark coloured and black earthworms have been found at 4,000 metres in Kashmir. The dark colours of many alpine insects are probably important for heat absorption, enabling the animals to warm up

Stonecrops (*Sedum*) are plants typical of rocky surroundings, growing where lack of water and harsh conditions create a hostile environment for vegetation (top). They are the food plant of the larvae of the beautiful Apollo butterflies (above), which, though characteristic of mountains throughout Europe, Asia and western North America, are most numerous and diverse in the Himalayas.

The brown bear is really an animal of the Palaearctic region but its range extends south to include the Himalayas. It is better able to withstand the cold bleak slopes above the tree line than the black bear which replaces it to the south. Like other bears it is an omnivorous feeder, taking vegetation, honey, fish and animals.

quickly during the early morning and complete their foraging before the ground becomes too dry. Some, like the woodlice and beetles, have developed highly reflective surfaces as protection against radiation.

The period during which this nival fauna can be active is severely limited, by the bitter cold at night and by strong insolation and low humidity during much of the day. Most insects, for instance, are active in the early morning, after the ground has warmed up slightly, or in the late evening or both. Correlated with the restricted feeding period and limited vegetation of this barren habitat, specialized feeding habits have developed. Only 30 per cent of the total Himalayan nival insect fauna feed on plants. Others feed on organic debris, scavenge on carrion or are active predators.

Insects are important alpine animals. Some species of butterflies, including 'whites' of the genus *Baltia*, mountain ringlets and the Apollo butterflies, are found flying and breeding at great heights. Apollo butterflies are related to swallowtails and their caterpillars feed on stonecrop. The pupae are not attached to a plant stem or leaf, as in most butterflies, but are enclosed in a cocoon on the ground; this is an adaptation for the harsh environment.

The majority of alpine insects, however, are wingless and even those that do possess wings seldom use them. In the Himalayas a flying insect above the tree line is in danger of being swept away by the stormy winds into a much less favourable environment and it may well be an advantage to be flightless. This cannot be the only answer, however, as some alpine insects of equatorial mountains where high winds are uncommon also lack wings.

Though winds are a problem for flying insects they are important in the maintenance of a rich invertebrate fauna close to the glaciers near the mountain peaks. Jumping spiders, for instance, have been found up to 7,000 metres. Though insects are common up to 5,000 metres, for a long time it was a mystery what these spiders could be eating. At last it was discovered that they were preying on anthomyiid flies in the open and, in cloudy weather, feeding on springtails beneath the rocks. Their prey feed mainly on fungus and rotting vegetation deposited as windblown debris at rock bases and cracks. It is interesting to note that the animals surviving in this inhospitable habitat and pioneers in zones beyond the plant life are members of the silverfish and springtail families, the oldest and most primitive of insects.

SOEPADMO

WARD

Pitcher plants are not confined to mountains but a great variety of them is found on Mount Kinabalu's forested slopes. The pitchers, which are modified leaves, are formed and coloured distinctively in the various species. They contain a liquid in which trapped insects are dissolved and digested.

Top: Kinabalu is one of the most isolated of the world's large mountains. It rises to 4,100 metres from a surrounding of rainforest which reaches almost to its summit and contains a number of animals found nowhere else in the world.

from lowland forest into the mossy montane forest of strange, bearded oaks and pine trees. Above 3,000 metres the vegetation is predominantly heathers, conifers, rhododendrons, mosses and lichens. This is the habitat of Kinabalu's famous array of exotic, carnivorous pitcher plants including the giant pitchers of *Nepenthes rajah*. The vegetation thins out till finally the bare rocky summit is reached.

Among the mammals found on Kinabalu but quite absent from the rain-forest are the short-tailed lesser moon rat, pencil-tailed mouse, the *Dendrogale* tree-shrew, four species of squirrels and, right in the alpine zone near the summit, numerous white-toothed shrews.

More obvious to the visitor, however, are Kinabalu's birds: the little cuckoo dove and the mountain minivet in the montane forests; a curious colourful crow, the treepie; the chestnut-headed minla, Borneo's only representative of a familiar Himalayan genus; and perhaps the most abundant birds of all, the laughing thrushes whose loud whistling calls brighten the morning mountain air.

At higher elevations mountain blackbirds and wren babblers are found. Different members of the same bird group have specialized to exploit different altitudinal vegetation zones. Thus the black-crested yellow bulbuls that are so common on the lower slopes of the mountain are replaced higher up by the pale-faced bulbul and the Bornean blue flycatcher is replaced by the white-fronted blue flycatcher. Finally, right up in the alpine zone, catching insects among the moss and heathers, is the conspicuous mountain blackeye.

Similar though their ecology may be in many respects, there are two important differences between the Himalayas and such equatorial mountains as Kinabalu. Since it lies near the equator, Kinabalu shows little seasonality of climate. Nights are cold and days are very warm but each month the weather is virtually the same. Since plant growth continues throughout the year there is a constant, if limited, food supply. Consequently animals can breed throughout the year and have no need to make seasonal migrations up and down the mountain in search of new pastures.

The second important difference is that isolated equatorial mountains are not whipped by the strong winds that scourge the larger temperate massifs. This greatly reduces the amount of wind-blown debris that is carried to the higher slopes and this in turn limits the arthropod fauna that can survive at these altitudes.

Mount Kinabalu

Apart from the big mountain ranges like the Himalayas there are other high peaks in the Oriental region, isolated peaks, rising like islands out of the rain-forests of Malaya, Sumatra, Java and Borneo. Many of these are volcanoes but the tallest of all is a single massive rock of igneous granite that rose up out of the ground about nine million years ago. Mount Kinabalu in north Borneo rises over 4,000 metres above sea level, almost twice as high as any other peak in Borneo.

Kinabalu supports a fauna quite distinct from that of the surrounding lowlands. Most of its montane species are found on other peaks but several such as the Kinabalu friendly warbler are found nowhere else in the world. This particular bird gains its name from its amazing fearlessness of humans and its habit of hopping about the feet of its infrequent visitors.

Like the Himalayas, Kinabalu's vegetation is zoned horizontally. The climber quickly passes

In the drier parts of tropical Asia, where water buffalo cannot be kept, the principal beast of burden is the ox. Here, in India, yoked pairs prepare fields for planting after the first monsoon rains.

In tropical Asia, perhaps more than in any other region, the ecology of the animals is inextricably linked with that of Man. Man has had a very long history in Asia, living at high densities. Not only has his presence had a profound influence on the distribution of animals, but his agricultural activities have even altered the climate of some areas. An account of the ecology of the Oriental region would not be complete without considering the effects that Man, the most successful of all animal species, has had on the local wildlife.

The history of Man in Asia

In 1890 a young Dutch anthropologist, Eugene Dubois, set out full of optimism to search for evidence of Man's ancestors in Java. Within a few months he reported that he had realized his ambition at Trinil where he had discovered the skull cap and thigh bone of a race of early Man, *Pithecanthropus erectus*. Modern anthropologists have now included Dubois' finds within our own genus as *Homo erectus*. Subsequent research has revealed even earlier hominids in the Djetis beds of Java and it is apparent that Man has been present in Asia for a considerable time, probably over a million years.

During the Pleistocene most of the Asiatic continent was glaciated but Indonesia enjoyed a tropical climate throughout the period, making it an attractive refuge for many animals and for Early Man. With the lowered sea levels there were no insuperable barriers to large mammals reaching the Greater Sunda Islands. *Homo erectus*, an upright, walking, tool-using, brainy man, must have reached Java about three-quarters of a million years ago, at the beginning of the Middle Pleistocene. Soon after this Man is known to have been living in parts of China, using crude stone tools and fire for cooking and warmth.

Throughout the Pleistocene *Homo erectus* was evolving larger brain size, from about 860 cc. in Java Man to 1200 cc. in the later Solo Man. (The average brain size of modern man is 1350 cc.) Parallel physical evolution was occurring simultaneously in Europe and Africa. Stone implements associated with Palaeolithic Man in Asia are very crude compared with those developed in Africa and Europe and some archaeologists, therefore, regard South-East Asia as a backwater of human cultural development. This is probably not the case. It seems likely that in Asia Man had a well-developed technology using bamboo and rotan canes rather

POLUNIN/NHPA

than stone and these have left no lasting remains.

Palaeolithic Man probably lived in small, self-sufficient groups, widely dispersed and highly mobile. Although he probably had little or no effect on his floral environment, he did prey on the larger animal species and probably hunted several of these to extinction. In Java alone bears, hyenas, giant pangolins, two species of elephant, rhinoceroses, tapirs, hippopotamus, several ungulates, siamangs and orang-utans all disappeared during Palaeolithic times. Most of these probably died out as a result of overhunting.

By the end of the Pleistocene successive waves of larger-brained hominids of our own species *Homo sapiens* spread across Asia. Using rafts and dugout canoes they reached Indonesia, Celebes and Australasia. *Homo erectus* disappeared to make way

Beautifully terraced paddy-fields or *sawahs* are the result of centuries of human labour. The cultivation of *sawah* rice is ecologically the most efficient and productive method of agriculture Man has yet discovered.

for the Mesolithic culture. First to arrive were the Negritos, short, dark, curly-headed men with broad flat noses. A second wave of better-equipped Palaeo-Melanesoids followed, forcing the Negritos back into the hills and forests, where their descendants can be found today among the aborigine tribes. Long-haired Veddahs spread through India and Ceylon and were later ousted by black-skinned Dravidians.

Mesolithic Man left kitchen middens of sea shells, flake-blade stone tools, horn and bone tools and, in some caves, the bones of prey animals and funeral relics. Here, too, the first signs of art and magic are to be found in the animal paintings and negative handprints embellishing the cave walls. From these remains and from observations of tribes from remote parts of southern Asia and Australia who still practise the same economy today, it is possible to reconstruct a picture of Mesolithic life in tropical Asia.

The people practised a hunting and gathering economy, the men travelling in bands after game while the women stayed nearer home collecting roots, fruits and shellfish. Although they did not practise agriculture they must have discovered, like the Veddahs and Australian aborigines today, that new yams and fruits grew where they had discarded shoots and seeds.

They probably lived in small nomadic groups wandering about within large tribal boundaries. Particularly good food sources, such as shellfish banks, may have supported permanent groups. The people sheltered in caves and rock shelters and wore bark cloth and decorative charms. They were able to fell trees with fire and burn out canoes. In arid areas they developed stone industries but in humid tropical regions they relied on a complex bamboo technology supplemented with rotan and hardwoods. Bamboo has a thousand and one uses, as containers for food and water, for cooking pots, fish traps and even for making fire by means of a bamboo saw. Men hunted with bows and arrows and fire-hardened bamboo spears. Bamboo blowpipes may have already been developed together with the use of vegetable poisons to kill large prey. To encourage more bamboo to grow the tribes burned and cleared the forest. Even at this primitive level Man was a threat to local wildlife, destroying the habitat and hunting the animals.

With the end of the Mesolithic period agriculture was developing in Asia. For a long time it was assumed that agriculture began in the fertile crescent of the Middle East and then spread to

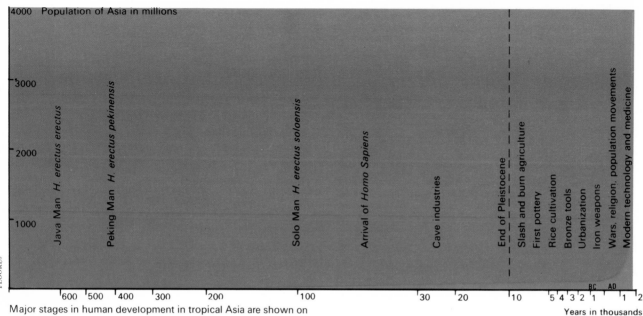

4000 Population of Asia in millions

3000

2000

1000

Java Man *H. erectus erectus*

Peking Man *H. erectus pekinensis*

Solo Man *H. erectus soloensis*

Arrival of *Homo Sapiens*

Cave industries

End of Pleistocene

Slash and burn agriculture

First pottery

Rice cultivation

Bronze tools

Urbanization

Iron weapons

Wars, religion, population movements

Modern technology and medicine

FLOOKES

600 500 400 300 200 100 30 20 10 5 4 3 2 1 BC AD 1 2

Years in thousands

Major stages in human development in tropical Asia are shown on a logarithmic time scale. The population has risen steadily with each cultural advance but in the last five hundred years has rocketed at an unprecedented rate.

other regions. Excavations in north Thailand, however, show that men in South-East Asia were cultivating peas, beans, cucumbers, water chest-nuts and rootcrops such as yams and taro twelve thousand years ago – three thousand years before the first evidence of agriculture in the Middle East. Pottery and, shortly afterwards, bronze tools made their first appearance in southern Asia. Clearly at this time the culture was advanced. Slash and burn shifting agriculture was practised and the transition into the Neolithic period was neither as culturally nor as climatically sudden as in other parts of the world.

With the Neolithic proper, about five thousand years ago, there was much more intensive destruction of virgin forest and intensive shifting agriculture on a communal village basis. Forest clearings were planted in the wet seasons with yams and taro. Taro requires irrigation and presumably terracing was developed at this time. Bananas and jackfruits were probably cultivated and the hill people began planting wild millet and hill rice. In the drier areas to the east of the Indonesian archipelago men fed extensively on the starch extracted from sago palms and in most coastal areas coconuts grew. Hunting of big game was probably less extensive

than during Mesolithic times and most protein was obtained by fishing and collecting shellfish. Wild jungle fowl, pigs and dogs were domesticated, as much for sacrificial offerings as for edible meat.

About 2000 BC large-scale movements of southern Mongoloid peoples spread through South-East Asia into Indonesia. They brought beautifully polished stone adzes, pottery and outrigger canoes. Undoubtedly they contributed enormously to the advances in agriculture in the archipelago and they had domesticated the water buffalo whose head became a decorative symbol still carved on the curved-roofed houses of Sumatra and Celebes today. These people were animistic, making offerings to numerous spirits that inhabited trees and streams. They worshipped the spirits of their ancestors and buried the dead in supine positions, enclosed within boat-like coffins with gifts for their spiritual journeys. Birds were regarded as spiritual animals and were incorporated in creation mythology and augury, as well as being used as decorative motifs. The cults of head-hunting and head preservation were practised and elements of this ritual can still be found in Nagaland, New Guinea and Borneo.

The development of lowland irrigated rice fields

The caves of Madhya Pradesh in northern India have been used for human habitation since prehistoric times. At Bhimbetka, paintings on cave walls date back to the Mesolothic. Some, like these, are more recent and depict hunting scenes and caparisoned horses about 2,000 years ago.

Young Indian girls wade through the wet paddy fields, planting out the young rice shoots. Rice forms the staple diet for most of the population of tropical Asia and has been domesticated from indigenous swamp grass.

and abundant cereal crops led to an enormous population expansion on the lowland plains and along the banks and estuaries of the large rivers of tropical Asia. Excess of manpower led to urbanization, armies, the establishment of kingdoms and wars. Threeway interchange of culture occurred among the warring kingdoms of south China, South-East Asia and India.

The last three thousand years have been a history of violence and conquest but also of trading and increasing technology. Bronze was replaced with iron. Aryans invaded India. Cows, sheep and goats together with the Hindu religion spread across Asia as far as Bali. In 500 BC Buddha and Confucius were teaching in India and China. In 326 BC Alexander the Great conquered the Punjab. Later the Emperor Asoka conquered most of the India subcontinent, then sent out Buddhist missionaries in every direction. Buddhism teaches a respect for animal life and as long ago as 242 BC Asoka appreciated the need for conservation and with his fifth pillar edict gave protection to fish, animals and forests.

Several hundred years later the dramatic expansion of Islam began and, within a century, the Muslims, with their abhorrence of wild and domestic pig meat, spread through Arabia, Persia and North Africa. During the Tan Dynasty Chinese traders were visiting Indonesia to buy spices, fragrant sandalwood and animal products, birds' nests, hornbill ivory, rhinoceros horn and colourful bird feathers.

In the thirteenth century Ghenghis Khan's Mongol hordes swept across Russia, China and into India, carving out a vast kingdom. Kublai Khan continued his grandfather's conquests and, in 1282, his Tartar bowmen destroyed the 'invulnerable' elephant cavalry of Burma. During the fourteenth century Tamurlane ravaged his way across north India and, in 1526, Babar began his Moslem invasion of India to set up the Mogul Empire. Arab sailing boats spread Islam right down into Indonesia, trading for the pepper and spices which were shipped back to Europe. The high cost of spices led in the sixteenth century to a wave of European exploration to the Orient and the beginnings of colonialism. Portuguese, Spanish, Dutch, French and English trading posts developed into colonies and empires over the whole of tropical Asia. Christianity was added to the growing medley of religions.

Colonialism encouraged new trades in the

Oriental region, such as the growing of opium, tea, coffee, rubber and cotton. These resulted in further clearance of natural vegetation and a consequent increase in population. The more recent advent of modern medicine has largely controlled disease so that it no longer operates as a natural check on numerical growth. The population of India remained steady at about 100 million from the third century BC until the seventeenth century AD. It then started to rise gradually before rocketing towards the 600 million of today. The exponential population growth of Java has been even more dramatic, involving a twenty-five-fold increase since the year 1800.

It is this huge unchecked explosion of mankind which now threatens the wildlife of tropical Asia. More and more land is required for agriculture to support these masses and serious overgrazing by cows, sheep and particularly goats has resulted in severe degradation of land in the more arid parts of the region. Roads and railways have opened up previously inaccessible areas and little natural forest will soon be left unless specifically protected.

Deforestation

Ever since men first adopted slash and burn agriculture, tracts of forest and woodland have been cleared for planting crops. The trees are felled and burned to provide ash for fertilizer and the land is planted with hill rice, tapioca, maize, chilli peppers and pawpaws. These produce good crops for a year or two but, as the soil becomes leached of nutrients, returns diminish. At this point the cultivator should move on to another patch and leave his old fields fallow for several years before clearing the vegetation again. With a high and expanding human population, however, the system breaks down. More and more virgin forest is cut and land is too scarce to be allowed to lie fallow and revert to its natural state. Instead the land is kept under crops for as long as possible until, eventually, coarse heath grass is all that it can support. Every year this grass is burnt to promote new growth for grazing for sheep and cattle and this, in turn, can lead to further degradation of the soil. Where human populations are high the effects on the habitat are greatest. India was once almost fully wooded but now less than 18 per cent of the subcontinent is forested. China has only 9 per cent of land still under forest. Large areas of heath 'blang' are now spreading in parts of Indonesia which once sported luxuriant rain-forest. The process can

be quite irreversible and the long-term consequences disastrous.

The forest preserves the soil and holds up water run-off. Flash storms falling on deforested areas run immediately into the rivers, causing catastrophic floods as well as washing away all the fine tilth. It is estimated that 35 tonnes of soil are washed from every 1,000 hectares of land in parts of the Ganges watershed each year. Deforestation can even result in climatic changes, with hot air thermals over cleared areas diverting rain clouds to cause artificial deserts.

More recently deforestation has speeded up for another reason: the enormous value now placed on timber. Vast areas are being felled every year. Since it would take hundreds of years for climax forest to regrow, reforestation schemes usually involve the planting of pine or eucalyptus, which form zoological deserts able to support virtually no indigenous wildlife.

Deforestation has undoubtedly destroyed the homes of millions of wild animals. Some species are already extinct and many more, like the orangutan, may become so. In many countries there is virtually no lowland rain-forest left. What remains on highlands is broken into small isolated blocks not large enough to support viable long-term populations of many species. Large carnivores, for instance, need a wide hunting range if they are not to overexploit their prey. They cannot survive in small stands of forest. With the disappearance of the predators a natural control is removed from the prey species so that they may breed unchecked until they literally eat themselves out of food.

Where tracts of forest are so broken up the whole food pyramid starts to erode. Even the delicate natural balance between plants and their pests begins to break down. Naturally the abundance of pest and plant are balanced. If the plant becomes abundant the numbers of its pests increase, causing a check on the plant. The abundance of the two species oscillate slightly out of phase. In a large population the pests will never become so common as to wipe out their specific plant; scarcity of food

Iban women watch as new clearings are burned for cultivation of hill rice in Borneo. In South-East Asia these slash and burn methods of agriculture have been practised almost unchanged for twelve thousand years.

will cause a gradual decline in their numbers before this occurs. In a small block of forest, however, many of the more uncommon plants will only be represented by a few individuals and the fine balance with their pests will not be maintained. The fluctuations of numbers will become wilder. Either the plant may become too scarce for the pest to survive or the pest might eat its host out completely so that neither species survives. One can predict with certainty that any small, isolated stands of forest that are left will, in the long-term, fail to preserve the diversity of life and the ecological complexity that is found in the vast expanses of lowland rain-forest. Sadly, it seems increasingly unlikely that any of these large tracts will survive.

Overgrazing

Overgrazing has caused terrible degradation of land in Asia as in other tropical regions. Far too many animals are kept on inadequate pasture and the result is severe soil erosion. When grazing becomes too poor to support cattle it can still support sheep and, when poorer still, goats. The idea that goats are benefactors, enabling men to subsist in areas they could not otherwise exploit is a dead-end concept. Goats will eat the last of the cover, climbing trees to consume the shoots and even bark. When the goats can no longer find food, nothing can.

Deforestation and overgrazing have led to the spread of deserts in historical times across north Africa, Arabia, Iran and into north-west India. Severe overgrazing within India has made most of that country agriculturally useless. Protected by religious taboos, cows have run wild causing severe damage in many areas. Between 1900 and 1940 the number of cows in India rose from about 84 million to 147 million whilst the amount of pasture available decreased by over 20 million hectares. Today there are more than 200 million cattle.

Apart from the degradation of the land, overgrazing has resulted in an enormous decline in the numbers of natural grazing and browsing ungulates. Large predators such as tigers and leopards have consequently become extremely scarce because they have few natural prey and are shot if they turn to taking domestic stocks.

A mixture of natural ungulates is far more efficient at producing meat from a given area of pasture than is domestic stock. Moreover land can support greater densities of natural ungulates, animals which have evolved within that environ-

POLUNIN/NHPA

ment, without suffering erosion. The ecological answer is clear. Domestic herds must be reduced till they are within the carrying capacity of the pasturage available and, where possible, natural herbivores should be ranched in preference to domestic breeds. As always, however, with human problems the answer is not so easy. The livelihoods, customs and religious and cultural beliefs of too many people are involved for any simple solution.

Introduced species

Man has caused great ecological changes in many parts of the world by intentional or inadvertent introductions of foreign species of plants and animals into new regions. Frequently these introductions are quite harmless and add to the diversity of local wildlife, as with the introduction of brown trout into the rivers of Kashmir, but sometimes they have resulted in such severe pest problems as the rabbit and the prickly pear cactus in Australia. Three examples are of some relevance to the ecology of tropical Asia.

The lantana weed, an ornamental, spiny branched bush, was introduced into the Pacific region from tropical America as a barrier to contain

Huge tree-grappling machinery helps speed up logging operations on the Kinabatangan river in Borneo. The forests of South-East Asia are being felled at a frightening rate. The process is irreversible as it would take hundreds of years for climax forest to regrow. The long term effects of extensive deforestation may well be disastrous.

cattle. The result has been disastrous as the plant has spread over whole pastures and rough grass-land throughout tropical Asia and Africa. It has ruined extensive grazing for domestic and wild animals alike.

The water hyacinth is a decorative water-plant with purple flowers and is native to tropical America. In 1894 the plant was cultivated in the famous Bogor Botanic Gardens in Java. It has since spread throughout Indonesia and the Philippines. By 1902 it was in Indo-China and by 1905 had swept right through India to Ceylon. This fast advance is due to its ability to reproduce vegetatively by growing new stolons, shoots which grow from its base. In a single year one plant may give rise to 100,000 others. The weed is a terrible pest, clogging irrigation ditches, hindering fishing and navigation and transforming fish spawning areas.

In 1847 the East African giant snail was released in the garden of the Bengal Asiatic Society. Within ninety years it had spread throughout tropical Asia to Indonesia. Although causing little damage in its native African forest this species has proved a real menace in Asia. It is a large snail with a tremendous rate of reproduction and is a voracious feeder. It climbs trees and attacks crops of tea, rubber, cocoa, bananas, citrus trees and ground cover as well as numerous wild plant species.

Overhunting

From Stone Age times, Man's hunting activities have led to the disappearance of many of his prey species. This is not merely an Asian phenomenon. The extinction of many genera of large mammals in North America, Africa and Madagascar coincides not with any particular climatic change, but with the arrival of the first human hunters. In predators that specialize predominantly on a single prey species there is a balance in the numbers of predator and prey. In those species that take a wide range of prey, as soon as one prey species becomes rare the predator switches its attention to alternatives. The difference with Man is that he is not entirely dependent on animal prey. He can survive on a diet with no animal protein. Even among the most carnivorous of human groups, the North American Indians, animal protein rarely exceeds 40 per cent of the total diet. Scarcity of prey does not cause human numbers to drop; instead it leads to an intensification of hunting effort to keep up supplies. This is well illustrated by the marine fishing industries, where every year more boats spend longer at sea but the catch continues to decline. Whereas natural carnivores are numerically scarce compared with their prey, men, whose numbers are boosted by other food sources, are often more abundant than the animals they hunt. The large, long-lived, slow-breeding mammals have been particularly susceptible to overhunting by Man.

This overhunting has continued into recent times and is probably responsible for the disappearance of the argus pheasant from many of its former haunts. Overzealous collecting of the eggs of leatherback turtles and uncontrolled hunting of dugongs have seriously endangered these species. Overfishing is also serious, and, in particular, the antisocial practices of killing fish with poison and dynamite have led to the virtual disappearance of large fish from many rivers.

Recently, however, hunting has been adopted for

reasons other than food. Sport in India is centuries old. Hawks for falconry and cheetahs for chasing blackbuck have been kept since the Mogul Empire. The Emperor Akbar is said to have kept a thousand trained cheetahs for hunting. The practice lasted into this century but today the Indian cheetah is extinct. Trophy hunting which was a European's pastime during colonial times became anybody's game after Independence, with a complete disregard for the numbers of animals shot. The populations of large game have been further depleted by irate villagers killing wild animals which have destroyed their crops or carried off their domestic stock. Tigers and leopards have become particularly scarce and the lion is now found only in the Gir Forest Reserve.

Hunting for elephant ivory, rhinoceros horns and the casques of the helmeted hornbill have resulted in the extremely local distribution of elephants and hornbills and the near extinction of the three Asian rhinoceros. Crocodiles have been so hunted for their valuable skins and eggs that they have died out through most of their range. Now the large-scale trapping of monkeys for medical research is seriously endangering the status of some species.

The changing scene

Man is actively reshaping the landscape. Some of his changes, such as the creation of reservoirs and the irrigation of arid regions, have been of advantage to wildlife and have helped enrich these areas but most of his plans have led to an impoverishment of the local fauna. The building of highways, draining of swamps and clearing of forest for agriculture and timber have all taken their toll.

Tropical Asia is a relatively underdeveloped region with too great a human population. There is too little food and too little land all round. The short-term need for development is painfully obvious, but without proper ecological planning such development may lead to long-term disaster. The thick vegetation cover provided by natural forest is vital for preserving the soil, fixing carbon dioxide from the atmosphere and releasing oxygen. The controlled level of slow silt run-off from forest ensures a long-term supply of fertile soil for lowland agriculture. Without the forest there will be a steady degradation of soil fertility. The use of fertilizers is only a short-term solution. Guano is in limited supply and artificial fertilizers made from oil are expensive and also dependent on limited resources of fossil fuels.

HARDING

ROSS/NSP

Although the flowers of *Lantana* are attractive to this Pierid butterfly, for Man the spread of this weed has been disastrous. A native of tropical America, *Lantana* was introduced to confine cattle but has spread unchecked to choke whole pastures and form impenetrable thickets.

Top: A familiar sight in Hindu India, cattle wander freely. Protected by religious taboo, cows can be a great nuisance, holding up traffic and overgrazing pastures.

Punans, natives of Borneo, still prefer traditional wooden blowpipes for hunting birds and monkeys. The 15 centimetre darts are tipped with poison made from the sap of the ipoh tree. Accuracy and silence make the blowpipe an ideal weapon for the thick rain-forest.

Nothing Man can plant in its place will be as luxuriant or act as efficiently in fixing solar energy as natural forest. Despite all Man's use of irrigation and chemicals, his monoculture crops are far less productive than natural vegetation. Too much sunlight is wasted as it falls on bare earth and crops are highly vulnerable to pests and disease. Maximum productivity is reached only when full, green active plant cover is presented to the sunlight throughout the year and the secret of this is species complexity. In natural forest different species of plants, with different seasonal phases, trap the sunlight at different horizontal levels and virtually nothing is wasted. Until Man can develop crops of comparable efficiency, there will be a need to preserve natural vegetation. To chop down all the forest for a quick sale of timber is a very short-sighted view. Logging should be controlled to a level within the productive capacity of the forest and human population growth must be checked so that pressure for land use is eased. Meanwhile Man has a lot to learn about the ecology of natural forests if he is to improve the productivity of his own crops. As a species we simply cannot afford to destroy the rain-forest until we know more about how it works.

Throughout his history Man has exploited animals, hunting wild prey, domesticating fowl, cattle, sheep, camels and elephants for food, clothing and transport or killing game for sport. Today we are learning to appreciate animals as an amenity for pleasure and the value of natural reserves for leisure and enjoyment will become increasingly recognized. As natural habitat becomes scarcer and the human population continues to expand, more and more people will want to visit these reserves. It is difficult for those who are hungry to appreciate and enjoy the beauties of wildlife but the Oriental region is potentially one of the wealthiest areas in the world and in time a much higher standard of living will be achieved there.

The establishment of a balance between natural and artificial environments must be to the long-term advantage of both Man and animals in tropical Asia. This is not a new idea; India had its first reserve as long ago as 300 BC and most of the countries of the Oriental region have established parks where people can see and appreciate the natural wildlife. While men as hunters, cultivators and timber users have been the enemies of the animal kingdom, men as tourists may well be the salvation of many of the threatened species.

Bibliography

ALI, SALIM: *The Book of Indian Birds*
Bombay Natural History Society, 1964

ATTENBOROUGH, D: *Zoo Quest for a Dragon*
Lutterworth Press, London, 1957

DWIGHT DAVIES, D: *Mammals of the Lowland Rain-forest of North Borneo*
Bulletin of the National Museum of Singapore No 31, 1962

DORST, J: *Before Nature Dies*
Collins, London, 1970
Houghton Mifflin, Boston, 1970

FOGDEN, M & P: *Animals and their Colours*
Peter Lowe, London, 1974
Crown Publishers, New York, 1974

GEE, E P: *The Wild Life of India*
Collins, London, 1964

HEEKEREN, H R VAN: *The Stone Age of Indonesia*
Martinus Nijhoff, The Hague, 1972

MACKINNON, J: *In Search of the Red Ape*
Collins, London, 1974
Holt, Rinehart and Winston, New York, 1974

MATTHEWS, S W: *The Changing Earth*
National Geographic Magazine, January, 1973

MEDWAY, LORD: *Wild Mammals of Malaya and Offshore Islands including Singapore*
Oxford University Press, Oxford, 1970

MORRIS, R & D: *Men and Pandas*
Hutchinson, London, 1966
McGraw-Hill, New York, 1967

RIPLEY, S DILLON: *The Land and Wildlife of Tropical Asia*
Life Nature Library, Time-Life International, 1964

SCHALLER, G: *Deer and the Tiger: a study of Wildlife in India*
University of Chicago Press, Chicago, 1967

SHELFORD, R W: *A Naturalist in Borneo*
Allen and Unwin, London, 1916

SMYTHIES, B E: *The Birds of Borneo*
Oliver and Boyd, London, 1960

TWEEDIE, M: *Animals of Southern Asia*
Paul Hamlyn, London, 1970

WALLACE, A R: *The Malay Archipelago*
Macmillan, London, 1869

Glossary

bioluminescence Light produced by living organisms.

biome Major habitat type.

brachiation Mode of locomotion used by gibbons and other apes, hanging beneath branches and using the arms to pull the body forwards.

brachyptery Reduction of wings.

browser An animal that feeds on leaves and shoots of bushes and trees.

canopy The upper layers of the forest, the leafy crowns of the trees.

chelicerate Invertebrate sub-phylum, including spiders, scorpions and king crabs.

climax forest End point of species succession with mature hardwoods dominating.

commensal relationship Two species living together for mutual benefit.

cryptic coloration Any coloration that helps to conceal an animal.

culms The stems of grass, reeds, bamboo etc.

detritus Decaying matter, either vegetable or animal.

drey A squirrel's nest of twigs.

dipterocarp A tree with two-winged fruits; most of the tall trees of the South-East Asian rain-forest are dipterocarps.

disruptive coloration Boldly patterned coloration that breaks up the shape of an animal and makes it less conspicuous.

ecological separation An arrangement whereby animals with similar habits avoid competition by developing preferences for different types of food, different habitats or different times of day.

ecosystem The living animals and plants of an area, their relationships with one another and with their environment.

ectoparasite A parasite that feeds externally on its host.

epiphyte A non-parasitic plant that grows on another.

eversible Able to be turned inside out.

generalized dentition Teeth not adapted for any special diet.

habitat Living space for organisms within the physical and biotic environment.

hectare Area equivalent to one hundred metres square.

hyphae Fungal root-like threads.

insolation Exposure to the sun's rays.

interglacial The warmer period between two ice ages.

interpluvial The drier period between two wet ones.

mandible The upper or lower part of a bird's bill.

melanic Having dark or black pigment.

metabolic water Water produced physiologically by an animal, an adaptation for desert living.

microhabitat A habitat found under stones, logs, in bark etc., which provides conditions different from those in the surrounding area.

niche An animal's place in the system of nature.

niche separation see **ecological separation.**

nival Living or growing in snow.

neotonous Remaining in larval form for the whole of life.

Palaearctic Major **zoogeographic** region comprising northern Asia and Europe.

patagium The wing-like membrane of a flying squirrel; a bat's wing.

photosynthesis The process by which green plants trap solar energy for the conversion of water and carbon dioxide into sugars.

pinnate feather Feather with a central quill and lateral radiating vanes.

plumose Feathery or plume-like.

pluvial A prolonged period of excessive wetness.

pelage An animal's coat or fur.

poikilotherm A cold-blooded animal whose temperature varies with that of the environment.

radiation Evolutionary divergence of species from a common generalized stock.

raptor A bird of prey.

relict species Surviving form of a previously more widespread type.

scansorial Adapted for climbing, using several strata of the forest.

tectonic Structural; here, concerning the structure of the earth.

ultrasonic Sound of high frequency, beyond the human spectrum of receptivity.

vibrissae Bristles such as cats' whiskers, which are sensitive to touch.

xerophytic Adapted for life in dry conditions.

zoogeographic regions Divisions of the world corresponding approximately to the five continental masses of North America, South America, Africa, Australasia and Eurasia, with the last divided into two regions, the **Palaearctic** to the north and the Oriental to the southeast. Originally based on the distribution of different bird species, each region has its characteristic fauna.

Index

Numbers in italics refer to illustrations.
† *indicates extinct.*